Sustainable Materials and Technology

Sustainable Materials and Technology (SMT) book series publishes research monographs (both edited and authored volumes) showcasing the latest developments in the field and comprehensively covering topics such as:

- Recycling of waste into useful material and their energy applications
- Catalytic action of Nano oxides for efficient carbon reforming process
- Sustainable technologies for plastic transformation
- Bifunctional nanoparticles for sustainable water splitting applications
- Sustainable dying and printing
- New materials from waste
- Sustainable Manure Management and Technology: Potentials, Uses and limitations
- Sustainable Mechanical Engineering Approach
- Sustainable biochemistry for the improvement of health
- Sustainable development of Mechanical recycling of automotive components
- Sustainable-waste recycling and conversion in useful materials for different applications
- Sustainable development of inexpensive Nano-photo catalysts
- Sustainable development of recycling of discarded lithium ion batteries
- Modern sustainable cement and concrete
- Sustainable adsorbent for hazardous removal
- Sustainable superior electromagnetic shielding materials
- Excellent sustainable nanostructured materials for energy storage device
- Sustainable development of heavy metal detoxification from water
- Carbon dioxide utilization for sustainable energy
- Sustainable development in green syntheses of materials
- Environment friendly and sustainable cloth for garments application
- Sustainable design and application of eco-materials
- Nanoparticles for sustainable environment applications
- Sustainable remediation of industrial contaminated water towards potential industrial applications
- Biomaterials for sustainable bioremediations

Anish Khan · Khalid A. Alzahrani
Editors

Microalgae Biotechnology and Sustainable Development

Springer

Editors
Anish Khan
Centre of Excellence for Advanced Materials Research
King Abdulaziz University
Jeddah, Saudi Arabia

Khalid A. Alzahrani
Department of Chemistry
Faculty of Science
King Abdulaziz University
Jeddah, Saudi Arabia

ISSN 2731-0426 ISSN 2731-0434 (electronic)
Sustainable Materials and Technology
ISBN 978-981-95-2596-6 ISBN 978-981-95-2597-3 (eBook)
https://doi.org/10.1007/978-981-95-2597-3

This Springer imprint is published by the registered company Springer Nature Singapore Pte Ltd.
The registered company address is: 152 Beach Road, #21-01/04 Gateway East, Singapore 189721, Singapore

Contents

Flashing Light Effect as a Stress to Maximize Microalgae Growth

Maya Margaritova Zaharieva, Fabiano Bisinella Scheufele, Alexander Dimitrov Kroumov, and Hristo Najdenski

Abstract A phenomenon called flashing light effect (FLE) has very solid theoretical basis, which lies in the intimate metabolic mechanisms of photosynthesis. Light is utilized instantly, but the syntheses of cell components are slow processes compared to that. Hence, FLE is crucial to increasing biomass quality and quantity. Having it as a tool in microalgal cultivation, methods are extremely beneficial. The frequency of light/dark cycles can be optimized according to cell pigments abilities of any particular strain. This can be done in modern photobioreactor (PBR) design where geometry is chosen on the basis of computing fluid dynamics simulations. Many scenarios of flow distributions are analyzed before choosing optimized frequency of light/dark cycles inside the vessel. Further, real active experiments to discriminate the competing hypotheses of the models must be planned and executed. This chapter analyzes in detail major advantages of growth when FLE is applied.

1 Introduction

The phenomenon of light stress can be evaluated through a new technique, flashing light. It is proven that FLE can enhance photosynthetic efficiency highly compared with constant light (under the same energy input). That effect reflects in enhanced microalgae-specific growth rate and, consequently, in higher biomass concentration levels in the photobioreactor (PBR) vessel. Techno-economical characteristics of the effect can be considered as well; it means FLE uses less energy when compared to constant light up to 80% depending on working conditions. Flashing light is one of the potential strategies. Flashing light is the intermittent light that can provide highly intense light from the microsecond time scale to longer time levels using

M. M. Zaharieva · F. B. Scheufele · A. D. Kroumov (✉) · H. Najdenski
Department of Infectious Microbiology, The Stephan Angeloff Institute of Microbiology, Bulgarian Academy of Sciences, Sofia, Bulgaria
e-mail: adkrumov@gmail.com

Engenharia de Bioprocessos e Biotecnologia—COEBB, Universidade Tecnológica Federal do Paraná—UTFPR—Campus Toledo, Toledo, Brazil

A. Khan and K. A. Alzahrani (eds.), *Microalgae Biotechnology and Sustainable Development*, Sustainable Materials and Technology,
https://doi.org/10.1007/978-981-95-2597-3_1

artificial pulse light source devices [52]. According to [22], by using proper flashing light, the phenomenon of photoinhibition can be minimized because light exposure is very short and is not able to cause damage to the photoreaction center. State of the art shows that FLE successfully was applied in autotrophic culturing of microalgal cells and depends on the quality of light-intensity, spectral distribution, and beam geometry of the light ray culture density. Special attention on that should be given on fundamental studies as reported elsewhere [6, 35, 48]. Parameters of PBR design and the flow distribution determined by mixing are of great interest, as well [3, 22]. The three main parameters characterizing the flashing cycles are light intensity, light frequency, and light/dark (L/D) cycles that is the fraction of time in which the microalgae are exposed to the light (Fig. 1).

These parameters define the range of light intensities to which the cells are exposed, as well as their frequencies. Therefore, the kinetics of mixing cycles varies greatly and changes between a millisecond time scale in algal PBRs to longer times by several orders [22]. It has been suggested that the alternation of L/D periods is beneficial to photosynthesis of microalgae.

The main characteristics of FLE include parameters such as flashing frequency (f), average light intensity Ia, and duty cycle (φ). They have been investigated for microalgal cultivation in the following ranges: 0.001–1 MHz, 2–1000 μmol m^{-2} s^{-1}, and 0.001–0.7, respectively [42, 45]. These works showed that for the given high-density culture (HDC), application of external flash can enhance light penetration in PBRs, and ensures high photosynthetic efficiency (PE) [1, 37]. In addition, the energy input can be reduced by up to 79% because of the FLE phenomenon [33].

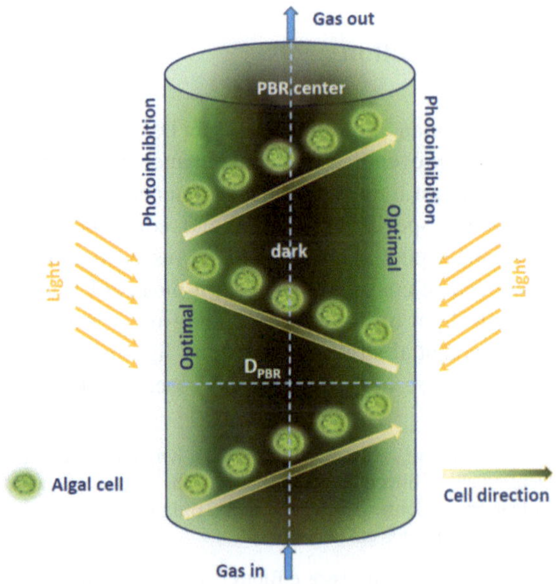

Fig. 1 Simplified scheme of the flashing light effects on microalgae cell culture during the mixing process

In order to give more understanding of how the light intensity effects the microalgal growth, we have to analyze microalgal physiology in PBR design. The best way to do that is to model the involved processes such as microalgal kinetics, mass transfer, and mixing conditions in the reactor vessel during CO_2 utilization. Microalgal kinetics is fundament of the strategy to build the complex mathematical description of the PBR as a system [19, 25, 28, 40, 41]. Hence, microalgal kinetics as a sub-system can be formalized in many steps to achieve the overall kinetic model. Such a fundamental description firstly was given by authors [11]. The model considered microalgae kinetics in photobioreactors (PBRs) and gives an understanding of how to analyze growth dynamics, photoinhibition, and recovery after that.

In Fig. 2, one may recognize the three-step process in autotrophic growth for the utilization of light by pigments. The first phase is a light-dependent stage where it is fixed as adenosine triphosphate (ATP) and nicotinamide adenine dinucleotide phosphate (NADPH). It must be noticed that in light reactions, water is hydrolyzed and oxygen is released as a product. Further, the cells utilize them in dark reactions for biomass needs of phase II (light-independent stage). In dark reactions, CO_2 is incorporated into the cell components (biomass) via a Calvin–Benson cycle. This phenomenon of light/dark (L/D) cycles has been used with great success when build innovative closed PBRs. Since the hydrodynamic conditions determine the utilization of light photons in such reactors, they must be precisely controlled [49, 50]. Further, the authors showed how to model microalgal kinetics in an actual tubular PBR.

By analyzing Fig. 2, one may see that two different regions can be distinguished in the volume of the PBR, one dark and one with illumination. Photons are taken by microalgae cells in the illuminated region by open photosynthetic factory (PSF) (Stage X1), where two options are possible (transfer to stage X2) and PSF deactivation

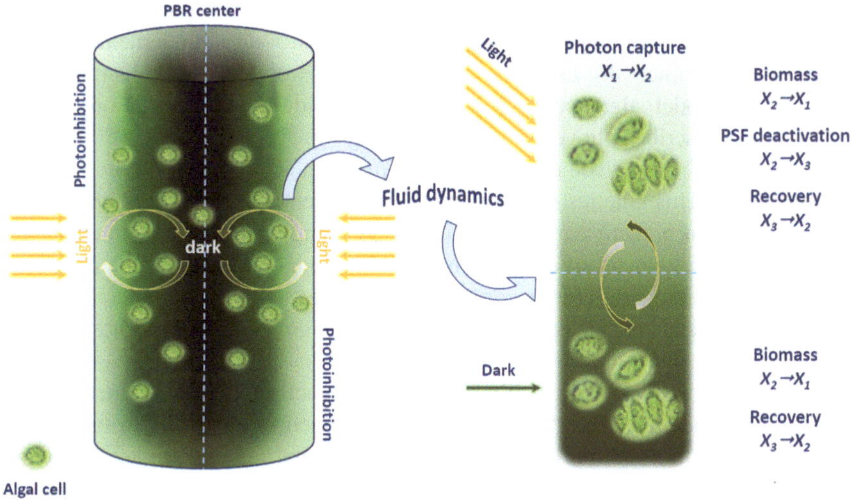

Fig. 2 Scheme of the light/dark cycle determined by fluid dynamics, photon capture, and growth of biomass

(transfer to stage X3). Hence, the cells in a loop procedure enter into to the dark zone, where PSF recovery takes place. This kinetic model [Eqs. 1–4] was realized experimentally, and its validity was verified by Wu and Merchuk [51].

The system can be written as follows:

$$\frac{dx_1}{dt} = -\alpha I x_1 + \gamma x_2 + \delta x_3 \tag{1}$$

$$\frac{dx_2}{dt} = \alpha I x_1 - \gamma x_2 - \beta x_2 \tag{2}$$

$$x_1 + x_2 + x_3 = 1 \tag{3}$$

$$\mu = k\gamma x_2 - Me \tag{4}$$

where I is the constant of the incident light illumination (I_0) [μE m^{-2} h^{-1}]; x_1, x_2, and x_3 are the fractions of PSF in the open state, closed state, and inhibited state, respectively [dimensionless]; α stands for the kinetic constant for x_1/x_2 [m^2 μE^{-1}]; β is the kinetic constant for x_2/x_3 [m^2 μE^{-1}]; γ is the kinetic constant for x_2/x_1 [1/h]; k stands for the kinetic constant for growth [μE m^{-2} h^{-1}]; μ stands for the specific growth rate [h^{-1}]; and Me stands for maintenance of the growth [h^{-1}]. More details about photosynthetic factory can be found elsewhere [26].

Light penetration and intensity of light on the wall and inside the PBR vessels are crucial criteria for any PBR design before considering the opportunity to optimize FLE and its frequencies. Many authors worked on this issue and showed the boundary conditions of PBRs configurations and dimensions which determined the reactor affectivity [5, 8, 46, 47]. Detailed investigation is done by the authors [34].

Analysis of FLE through the model equations (Eqs. 1–4) is fundamental theoretically and experimentally useful by controlling steps of cell metabolism and their rates. In the model, algae cells are subjected to incident light irradiation (I_0), and four phases (described above) of cell functioning were recognized [49, 50].

As the key characteristic of the cell in the model it is a result of the sum of all of the processes. This kinetics was validated for different closed PBRs with different L/D cycles. Hence, there are no limitations for the application of this kinetic model for process development of different microalgal strains of interest when culturing on industrial waste gases (for example, flue gas). The changes in the specific growth rate (SGR) depend on state parameters and PBR design which in any particular case must be analyzed in detail.

The scale-up effect on light illumination is a function of PBR diameter. When the diameter increases, the flow distributions changed dramatically in time which requires application of requires the application of internal baffles to organize the flows in the way to ensure an optimal FLE inside the vessel. Therefore, the mathematical modeling of the PBR as a complex system and its scaling up by using such a tool is unavoidable. Another issue that has to be considered is how to analyze the cell performance, on the macro- (population) and micro-(molecular) levels. Moreover, the interactions between the levels cannot be neglected and must be analyzed. When analyzing at the micro level, the light utilization must be considered in cells of microalgae pigments in chloroplasts [18]. The identification of pigments can be useful to identify their light absorption spectra which are characteristics of light source and material applied for closed PBR columns. Details about classification of microalgae pigments and their distribution in chloroplast can be found elsewhere [18, 20]. The energy transfer pathway is described by [30]. Absorption spectrum and wavelength range associated with microalgae pigments can be found elsewhere [15, 18, 30]. Physiological stress of microalgae cells under high light intensity can be studied for the high-density culture conditions. Theoretical analysis of the adaptation process is published by the authors [15, 30], and changes of cell composition are studied in detail by Lee and Palsson [29]. The process of photon intake and its efficiency of microalgal cells can be theoretically predicted. Light penetration into PBR volume depends on the biomass density, which can be described by Lambert Beer law. From the model followed at low cell density, light-intensity requirements must be low. Therefore, under higher biomass concentrations, the requirements for light intensity are high because of the mutual shading of the cells in the liquid. Hence, the PBR design and definition of working conditions, to determine key parameter of L/D cycle [14, 21, 38] are function of light supply under high biomass concentration. Design of one geometry of closed PBR such as a flat plate is well studied and published by authors [43].

Everything connected with light illumination of PBRs started with evaluation of their surface-to-volume ratio (S_f/V_{PBR}). This ratio plays a crucial role not only for light penetration but also to determine the strategy of how to control the FLE in the reactor vessel. Mathematical formulation of (S_f/V_{PBR}) and evaluation of tubular PBRs can be found in many works cited by [24]. The authors for the first time connected the PBR geometry with microalgae kinetics and theoretically analyzed the potential of different PBR columns to achieve high-density culture conditions.

Light criterion taking into consideration of S_f/V_{PBR} can be represented by Eq. (5):

$$\frac{S_f}{V_{PBR}} = \frac{\pi D_{PBR} H}{\pi H (D_{PBR}^2/4)} = \frac{4}{D_{PBR}} \tag{5}$$

$$\mathrm{Sf/V} = (\pi.\mathrm{D_{PBR}}.\mathrm{H})/(\mathrm{H}.\pi.(\mathrm{D_{PBR}}2/4)) = 4/\mathrm{D_{PBR}}.$$

If the D_{PBR} is equal to 1 cm, then $S_f/V_{PBR} = 4/D_{PBR} = 4/(0.01m) = 400m^{-1}$.

This simple model showed that light availability depends only on tubular PBR's diameter. PBR designs result in a maximum value of $(S_f/V)_{max} = 400\ m^2/m^3$ which corresponds to 1 cm diameter of the tubular reactor. The authors made their analysis of microalgal kinetics by linking the S_f/V with Monod simple model and achieved the form of specific growth rate as a function of PBR geometry and microalgae physiology. Building the biomass balance, the authors determined the effectiveness of tubular PBRs with different diameters. The conclusions from their work [24] are as follows:

- for low light intensities (not summer conditions), tubular PBRs with $D_{PBR} > 40$ cm are not effective because they achieved low biomass concentration about 0.5–0.8 g/L;
- for high light intensities (*e.g.*, I_0=3270.0 µmol m^{-2} s^{-1}), namely hot summers, tropical regions, and deserts, column PBRs with diameters up to $D_{PBR} = 20$ cm can be successfully used;
- tubular PBRs with diameters 5 cm to 15 cm have great theoretical potential in tropical regions (USA deserts, North Africa, etc.);
- Algae strains resistant to shear stress and having high SGR may successfully be used in tubular PBRs where biomass concentration up to 6 kg m^{-3} can be achieved.

It must be noted that 20 kg m^{-3} biomass concentration during photoautotrophic growth reported in the literature is possible to be achieved only in very thin 0.01 m (or less) tubular PBRs which is unrealistic for industrial applications (simulation not shown).

Therefore, the self-shading effect is a major problem in PBRs if hydrodynamics does not ensure optimal FLE depending on radial flow distribution of cells from light to dark zones of the vessel. It must be noticed that reduction of self-shading can be controlled and optimized by applying LED technology and its innovative equipment implemented inside the PBRs resulting in reliable high-frequency pulsed light. On the other hand, in continuous operation residence time of the cells plays a crucial role, as well.

Mathematical modeling by using computational fluid dynamics (CFD) software and application of innovative PBR design is able to determine boundaries of application of FLE parameters and benefits of columns and flat plate PBRs. [36] proposed an approach to link FLE and hydrodynamics in a new PBR so-called inclined wavy-bottomed photobioreactor (WBP). This PBR design allow the authors to analyze the relationship between process design and the flashing light effect in the chosen WBP configuration. The FLE is evaluated through the frequency spectrum of the variation of perceived irradiance with time. They claim that never before such research were done for characterization of flashing performance of PBR hydrodynamics. To calculate this time-varying system they developed and utilized a model that couples a radiation transport model and hydrodynamic transport. The authors used software ANSYS FLUENT® (Ansys Inc., Canonsburg, PA, USA) to solve the Navier–Stokes equations and the additional transport equations. The model was validated experimentally. All the details of the experimental setup, analysis, and numerical calculations can be found in their materials and methods [36]. The results obtained in this work highlights the crucial importance of such methodology to solve scale up problems of different types of real PBRs under high natural/artificial light illumination (1000–2000 µmol photons m^{-2} s^{-1}). It must be noticed that the modeling considers population level of knowledge about microalgal kinetics coupled with hydrodynamics and FLE in the particular PBR design. Further, the authors demonstrated that the spectral performance (3 Hz) of the inclined WBP was above that achieved by vertical columns PBRs and flat panel PBRs with similar optical thickness.

The main conclusion of this work is that the robust evaluation of any PBR performance depends on application of mathematical tools and analytical methods allowing quantitatively to determine the boundaries of benefits from FLE in real big scale required for maicroalgal cultivation and to prove the optimal geometry of PBR, which guarantees optimal hydrodynamic conditions.

The incorporation of multiphase CFD software was used earlier to simulate and analyze PBR hydrodynamics and experimentally proved its effectiveness [19].

Experimental verification of benefits from application of FLE is crucial from an economical point of view. The authors made an attempt to show it in their research [37]. Their work is especially important because of real practical application for the most common case of achieving nigh density culture condition in PBRs. Moreover, the chosen microalgal strain of *Chlorella* is widely used in industrial scale of microalgal biomass production. The study follows a strategy to investigate low-, medium-, and high frequency of FLE in dynamically growing Chlorella strain (*Chlorella kessleri* (UTEX 398). Application of high-frequency flashing light (37 kHz) showed an increasing in biomass concentration approximately 20% than that under continuous light illumination. The results from this work determined a very important practical application of this technique which is the application of high-intensity flashing light with low duty cycle in order to minimize mutual shading of cells. The authors proved that instantaneous photosynthetic photon flux (PPF) can be 10 times greater than the one maintained during continuous light illumination for the given 10% duty cycle. Therefore, FLE definitely will increase the functioning of a PBR under the high-density culture conditions. The authors claimed that the work of PBR can be controlled at the optimized level without performing costly genetic modification of microalgal cells and redirecting the phenomenon of photosynthesis or its mechanisms. Hence, the application of only one control parameter may play a crucial role in the optimization procedure of the PBR system.

Moreover, practical evidence of importance of FLE lied to deep theoretical investigations by using more sophisticated models. Theoretical consideration of photoinhibition and metabolic activity of microalgal cells as a response of light irradiance are bases of the development and application of useful mechanistic models in microalgal technology. Some recent attempts in the state of the art in the are reported in the literature [13, 17]. In the work of [17], the author made an analysis of the existing efforts of modelers to describe the system of photosynthesis-irradiance and response of the algal cells by showing the advantages and disadvantages of the results on the subject. Having the knowledge database in the area, they offered a model for photosynthetic unit (PSU). His results are useful and represent one opportunity to compare and fit photosynthetic rate to experimental data. Evidently, its model uses a classical equation first proposed by [16] for enzyme reactions that are inhibited by a high concentration of substrate, and the representation of the dynamics of light based on [23]. His findings on the photosynthetic response of phytoplankton to irradiance are formalized in a practical simple model based on the assumption of how the PSU and PSII worked under the given conditions. The author claims that his model is robust and may have wild applicability.

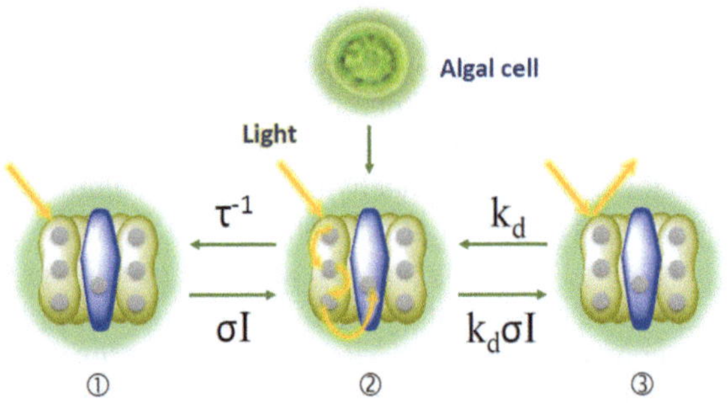

Fig. 3 Schematic representation of the three phases of reaction centers of PSII

Further achievement in theoretical biology on FLE is given in the work of authors [13], in which their postulates are based on the research of [17]. FLE is so important to be modeled because under certain applied and controlled boundaries, it may decrease photoinhibition and hence increase the biomass density in real PBRs. Many attempts to understand and use the benefits of FLE were done [1, 12, 42]. The researchers must be careful when comparing FLE with constant light conditions because the same average irradiance is used as a criterion [2].

In the work of researchers [13], most probable real situations of different light regime (the high/low flashing light) were considered, modeled, and analyzed. Such a phenomenon may likely take place in real conditions for high-density cultures (HDC) of microalgae where light is absorbed and scattered by the cells and its penetration is restricted to a minimum in the reactor vessel [19, 25, 27, 44]. As a consequence, cells oscillate randomly between the surface layer at high irradiance and the deeper layers at lower irradiance [9, 25].

The model of [13] considers three stages (1, 2, 3) of reaction centers of PSII same as in the model of [17]. Schematic representation of these phases can be drawn as follows (Fig. 3).

In Fig. 3, k_r stands for recovery rate [s^{-1}]; k_d is damage rate [–]; τ is turnover time [s]; σ is effective cross-section [m^2 μmol^{-1}]; I is light intensity [μmolm^{-2} s^{-1}]; P_{max} is maximum photosynthetic rate; I_{opt} stands for optimal light irradiance [μmolm^{-2} s^{-1}]. The authors simulate with the model scenarios how specific growth rate change for different I_{min}, I_{max} conditions and provided benefits of using FLE comparing to continuous light irradiation. Most of the findings are valid when model parameters are changing in some range (σ). It has to be noticed that in this work the model does not evaluate mixing. Hence, an extension can be realized when the flow distribution as a function of the hydrodynamic conditions will take place in the model. Therefore, the internal space of PBR can be divided in two or more layers with different light irradiation–illumination and light limitation. This will allow to calculate the average specific growth rate of microalgae biomass.

Conclusions from the theoretical work from [13] provide knowledge about the meaning of their large-T model which can increase the growth rate when compared to the continuous light for the given low local optical depth. In this case, combining a high inhibiting irradiance and a lower irradiance with a period greater than $1/k_r$ (in the range of hours) in average, the growth rate will be higher than the one considering continuous light regime with the same average irradiance.

Therefore, after the theoretical proof of phenomena involved in the intimate mechanisms of biochemical, photosynthetic, and other cell processes, the mathematical models are applied in all life sciences with great success because of the necessity to optimize the working conditions and state parameters. In microalgal cultivation biomass production and utilization of energy have to be maximized under the consideration of FLE [10, 36]. A challenging opportunity is to use a key tool for this goal which can be involved in the modeling procedure as well as in study state conditions [10].

Pulse light (PL) can be the case where many of its variables, such as frequency, duty cycle, integral light, light and dark period, and pulse intensity, have to be taken into account monitored and modeled. Some theoretical principles established in microbiology and microalgology about the modeling on population level must be highlighted. Using analogy, models describing cell kinetics and growth of microalgal cells are well known as empirical and mechanistic ones. The former does not consider the biochemistry of photosynthesis. Hence, kinetic parameters in these models are most often not linked to the biological phenomena of photosynthesis. Therefore, their predictive power is limited. On the contrary, mechanistic models have biological meaning and represent phenomena of the photosynthesis which give them more space of robust applicability.

Mechanistic models can in turn be divided into two main groups: "physiological models" and "state models." The former attempts to describe the photosynthetic process with extreme.

detail, thus requiring a lot of computational effort, since many variables and parameters are involved [4]. State models instead describe the photosynthetic process by splitting it into its main steps. These models are developed based on the concept of photosynthetic unit (PSU), defined as the sum of the light harvesting complex, the reaction center, and the associated apparatus. State models are generally preferred to physiological ones since they are more suitable for simulation purposes. In addition, PSUs can be monitored and validated through pigment and fluorescence measurements [4, 51].

Recent, development of state of the art where high-intensity and high-frequency pulsed light conditions are applied can be found in the work of scientists [39]. Especially important and crucially valuable are application of the knowledge about FLE and its boundaries in the field of wastewater treatment technologies [32]. The authors claimed that this is the first study of its kind applied to the photoheterotrophic organisms. They studied the system of purple non-sulfur bacteria (PNSB) in wastewater and planed experiments with artificial medium in order to perform the research. This case is valuable because of PNSB growth in wastewater (WW) with a high organic load. Therefore, utilization of PNSB for purification of WW received much attention [7]. The authors in a step-by-step experimental procedure studied and compared the difference between autotrophic microalgae under FLE and PNSB. Their research covers changes of biomass, protein, carotenoid, and bacteriochlorophyll concentration when applying flashing light stress. The most important parameters were monitored, including incident light (I_a), duty cycle (φ), and light frequency (f). Conclusions from their work highlighted the direction of utilization of FLE and condition how to reduce energy consumption compared with constant light application. By changing and monitoring the state parameters I_a, φ, and f, authors quantitatively achieved increase of biomass, protein, and pigment synthesis from 20% up to 60%, accordingly. Therefore, application of PNSB in WW treatment under controlled FLE provides potential for cost-effective PNSB-based technology in the market.

2 Future Prospects of Application of FLE in Microalgal Technology

Theoretical achievements and experimental verifications of FLE for particular PBRs and density culture conditions showed great potential in using pulse light to improve the overall performance of cultivation procedure. The modern and very powerful computer-aided engineering/computational fluid dynamics (CAE/CFD) software allows to simulate the hydrodynamics in any kind of PBRs, and it helps to fully understand how kinetics of microalgal growth and light conditions are linked. Evidently, mathematical modeling as a tool needs continued development. Appropriate application of the FLE key parameter and its high frequency and high intensity may minimize cost and time efforts to achieve maximum productivity by avoiding any genetic manipulation of microalgal strain. Further, on the basis of its application, PBR design can be greatly improved and made effective. Moreover, any new equipment that may give new possibilities to improve monitoring and control of FLE is needed for optimal and robust overall microalgal system performance. Applying knowledge about the potential of FLE in WW treatment widens the horizons of ecological and biotechnological perspectives for minimizing climate change effects and will help to restrict the spread of pandemics.

3 Conclusions

State of the art in the area of application of flashing light as a tool to improve the PBR system performance results in higher biomass concentration, photosynthesis rate, and pigment concentrations as well many secondary metabolites. The phenomenon of FLE definitely decreases the mutual shading of cells because of the control of their trajectories by PBR hydrodynamics. It must be noticed that its effect is extremely beneficial in high-density cultures taking place in industrial-scale conditions. The boundaries of high inhibiting irradiance and a lower irradiance must be carefully determined in FLE environment in order to achieve higher growth rate than the one in continuous light regime with the same average irradiance. Especially valuable can be FLE during production of high-value metabolites for pharmacy, medicine, and infectious microbiology among other commercial and scientific goals. Last but not the least, the mathematical tools used to estimate the FLE must be continuously improved and applied in all stages of microalgal technology development.

From an economical point of view, FLE can highlight the road to the formulation of optimal operation strategies for PBRs under high-frequency and high-intensity pulsed light while minimizing the energy waste and hence improving cost-effectiveness of the microalgal upstream processes.

Acknowledgements This work was supported by Grant КП-06-Н53/2 (11.11.2021) from Bulgarian National Science Fund which is gratefully acknowledge.

References

1. Abu-Ghosh S, Fixler D, Dubinsky Z, Iluz D (2016) Flashing light in microalgae biotechnology. Biores Technol 203:357–363. https://doi.org/10.1016/j.biortech.2015.12.057
2. Abu-Ghosh S, Fixler D, Dubinsky Z, Solovchenko A, Zigman M, Yehoshua Y, Iluz D (2015) Flashing light enhancement of photosynthesis and growth occurs when photochemistry and photoprotection are balanced in *Dunaliella salina*. Eur J Phycol 50:469–480. https://doi.org/10.1080/09670262.2015.1069404
3. Baroukh C, Muñoz-Tamayo R, Bernard O, Steyer J-P (2015) Mathematical modeling of unicellular microalgae and cyanobacteria metabolism for biofuel production. Curr Opin Biotechnol 33:198–205. https://doi.org/10.1016/j.copbio.2015.03.002
4. Bernardi A, Perin G, Sforza E, Galvanin F, Morosinotto T, Bezzo F (2014) An identifiable state model to describe light intensity influence on microalgae growth. Ind Eng Chem Res 53:6738–6749. https://doi.org/10.1021/ie500523z
5. Carlozzi P (2000) Hydrodynamic aspects and Arthrospira growth in two outdoor tubular undulating row photobioreactors. Appl Microbiol Biotechnol 54:14–22. https://doi.org/10.1007/s002530000355
6. Chang RL, Ghamsari L, Manichaikul A, Hom EFY, Balaji S, Fu W, Shen Y, Hao T, Palsson BØ, Salehi-Ashtiani K, Papin JA (2011) Metabolic network reconstruction of *Chlamydomonas* offers insight into light-driven algal metabolism. Mol Syst Biol 7:518. https://doi.org/10.1038/msb.2011.52
7. Chen J, Wei J, Ma C, Yang Z, Li Z, Yang X, Wang M, Zhang H, Hu J, Zhang C (2020) Photosynthetic bacteria-based technology is a potential alternative to meet sustainable wastewater treatment requirement? Environ Int 137:105417. https://doi.org/10.1016/j.envint.2019.105417
8. Da Silva MF, Casazza AA, Ferrari PF, Perego P, Bezerra RP, Converti A, Porto ALF (2016) A new bioenergetic and thermodynamic approach to batch photoautotrophic growth of Arthrospira (Spirulina) platensis in different photobioreactors and under different light conditions. Biores Technol 207:220–228. https://doi.org/10.1016/j.biortech.2016.01.128
9. Demory D, Combe C, Hartmann P, Talec A, Pruvost E, Hamouda R, Souillé F, Lamare P-O, Bristeau M-O, Sainte-Marie J, Rabouille S, Mairet F, Sciandra A, Bernard O (2018) How do microalgae perceive light in a high-rate pond? Towards more realistic Lagrangian experiments. R Soc Open Sci 5:180523. https://doi.org/10.1098/rsos.180523
10. Diotto D, Barbera E, Borella L, Trivellin N, Sforza E (2022) Modeling approach to capture the effect of high frequency flashing light in steady-state microalgae cultures. Ind Eng Chem Res 61:18258–18269. https://doi.org/10.1021/acs.iecr.2c03258

11. Eilers PHC, Peeters JCH (1988) A model for the relationship between light intensity and the rate of photosynthesis in phytoplankton. Ecol Model 42:199–215. https://doi.org/10.1016/0304-3800(88)90057-9
12. Fernández-Sevilla JM, Brindley C, Jiménez-Ruíz N, Acién FG (2018) A simple equation to quantify the effect of frequency of light/dark cycles on the photosynthetic response of microalgae under intermittent light. Algal Res 35:479–487. https://doi.org/10.1016/j.algal.2018.09.026
13. Fierro Ulloa JI, Lu L-D, Bernard O (2023) Theoretical growth rate of microalgae under high/low-flashing light. J Math Biol 86:48. https://doi.org/10.1007/s00285-023-01871-2
14. García-Camacho F, Sánchez-Mirón A, Molina-Grima E, Camacho-Rubio F, Merchuck JC (2012) A mechanistic model of photosynthesis in microalgae including photoacclimation dynamics. J Theor Biol 304:1–15. https://doi.org/10.1016/j.jtbi.2012.03.021
15. Graham L, Wilcox L (2000) Chapter 6: cyanobacteria. Algae. Prentice-Hall Inc., Upper Saddle River, pp 106–117
16. Haldane JBS (1930) Enzymes. Longmans, Green, New York
17. Han B-P (2002) A mechanistic model of algal photoinhibition induced by photodamage to photosystem-II. J Theor Biol 214:519–527. https://doi.org/10.1006/jtbi.2001.2468
18. Heldt H-W, Piechulla B (2011) The use of energy from sunlight by photosynthesis is the basis of life on earth. In: Plant biochemistry. Elsevier, pp 43–64. https://doi.org/10.1016/B978-0-12-384986-1.00002-8
19. Hinterholz CL, Trigueros DEG, Módenes AN, Borba CE, Scheufele FB, Schuelter AR, Kroumov AD (2019) Computational fluid dynamics applied for the improvement of a flat-plate photobioreactor towards high-density microalgae cultures. Biochem Eng J 151:107257. https://doi.org/10.1016/j.bej.2019.107257
20. van den Hoek C, Mann DG, Jahns HM (1995) Algae: an introduction to phycology. Cambridge University Press, Cambridge
21. Hu Q, Kurano N, Kawachi M, Iwasaki I, Miyachi S (1998) Ultrahigh-cell-density culture of a marine green alga Chlorococcum littorale in a flat-plate photobioreactor. Appl Microbiol Biotechnol 49:655–662. https://doi.org/10.1007/s002530051228
22. Iluz D, Alexandrovich I, Dubinsky Z (2012) The enhancement of photosynthesis by fluctuating light. In: Najafpour M (Ed) Artificial photosynthesis. InTech. https://doi.org/10.5772/31040
23. Kok B (1956) On the inhibition of photosynthesis by intense light. Biochem Biophys Acta 21:234–244. https://doi.org/10.1016/0006-3002(56)90003-8
24. Kroumov A, Gacheva G, Iliev I, Alexandrov S, Pilarski P, Petkov G (2013) Analysis of Sf/V ratio of photobioreactors linked with algal physiology. Genet Plant Physiol 3:55–64
25. Kroumov AD, Módenes AN, Trigueros DEG, Espinoza-Quiñones FR, Borba CE, Scheufele FB, Hinterholz CL (2016) A systems approach for CO2 fixation from flue gas by microalgae—theory review. Process Biochem 51:1817–1832. https://doi.org/10.1016/j.procbio.2016.05.019
26. Kroumov AD, Scheufele FB, Trigueros DEG, Modenes AN, Zaharieva M, Najdenski H (2017) Modeling and technoeconomic analysis of algae for bioenergy and coproducts. In: Algal green chemistry. Elsevier, pp 201–241. https://doi.org/10.1016/B978-0-444-63784-0.00011-4
27. Kroumov AD, Scheufele FB, Zaharieva MM, Zheleva-Dimitrova D, Najdenski H (2021a) Optimal biomass production by cyanobacteria, mathematical evaluation, and improvements in the light of biorefinery concept. In: Rastogi RP (ed) Ecophysiology and biochemistry of cyanobacteria. Springer Nature Singapore, Singapore, pp 401–429. https://doi.org/10.1007/978-981-16-4873-1_18
28. Kroumov AD, Zaharieva MM, Scheufele FB, Balabanova V, NajdenskiH (2021b) Engineering challenges of carbon dioxide capture and sequestration by cyanobacteria. In: Rastogi RP (ed) Ecophysiology and biochemistry of cyanobacteria. Springer Nature Singapore, Singapore, pp 351–372. https://doi.org/10.1007/978-981-16-4873-1_16
29. Lee C, Palsson BØ (1996) Photoacclimation of *Chlorella vulgaris* to red light from light-emitting diodes leads to autospore release following each cellular division. Biotechnol Prog 12:249–256. https://doi.org/10.1021/bp950084t

30. Lee RE (2008) Basic characteristics of the algae. In: Phycology. Cambridge University Press, pp 3–30
31. Lee Y-K, Pirt SJ (1981) Energetics of photosynthetic algal growth: influence of intermittent illumination in short (40 s) cycles. Microbiology 124:43–52. https://doi.org/10.1099/00221287-124-1-43
32. Lu H, Zhao R, Wang C, Zhang G, Chen C, Li B, Han T (2022) Exploration of flashing light interaction effect on improving biomass, protein, and pigments production in photosynthetic bacteria wastewater treatment. J Clean Prod 348:131304. https://doi.org/10.1016/j.jclepro.2022.131304
33. Lunka AA, Bayless DJ (2013) Effects of flashing light-emitting diodes on algal biomass productivity. J Appl Phycol 25:1679–1685. https://doi.org/10.1007/s10811-013-0044-1
34. Morita M, Watanabe Y, Saiki H (2000) Investigation of photobioreactor design for enhancing the photosynthetic productivity of microalgae. Biotechnol Bioeng 69:693–698. https://doi.org/10.1002/1097-0290(20000920)69:6%3c693::AID-BIT14%3e3.0.CO;2-0
35. Nogales J, Gudmundsson S, Knight EM, Palsson BO, Thiele I (2012) Detailing the optimality of photosynthesis in cyanobacteria through systems biology analysis. Proc Natl Acad Sci USA 109:2678–2683. https://doi.org/10.1073/pnas.1117907109
36. Olivieri G, Moroni M, Janssen M, Piersanti L, Mezza D, Bravi M (2021) Model-based prediction of perceived light flashing in recirculated inclined wavy-bottomed photobioreactors. Processes 9:1158. https://doi.org/10.3390/pr9071158
37. Park K-H, Lee C-G (2000) Optimization of algal photobioreactors using flashing lights. Biotechnol Bioprocess Eng 5:186–190. https://doi.org/10.1007/BF02936592
38. Richmond A (2004) Principles for attaining maximal microalgal productivity in photobioreactors: an overview. Hydrobiologia 512:33–37. https://doi.org/10.1023/B:HYDR.0000020365.06145.36
39. Saccardo A, Porcelli A, Borella L, Sforza E, Bezzo F (2024) Model-based optimisation of microalgae growth under high-intensity and high-frequency pulsed light conditions. J Clean Prod 469:143238. https://doi.org/10.1016/j.jclepro.2024.143238
40. Scheufele FB, Hinterholz CL, Zaharieva MM, Najdenski HM, Módenes AN, Trigueros DEG, Borba CE, Espinoza-Quiñones FR, Kroumov AD (2019) Complex mathematical analysis of photobioreactor system. Eng Life Sci 19:844–859. https://doi.org/10.1002/elsc.201800044
41. Schuelter AR, Kroumov AD, Hinterholz CL, Fiorini A, Trigueros DEG, Vendruscolo EG, Zaharieva MM, Módenes AN (2019) Isolation and identification of new microalgae strains with antibacterial activity on food-borne pathogens. Engineering approach to optimize synthesis of desired metabolites. Biochem Eng J 144:28–39. https://doi.org/10.1016/j.bej.2019.01.007
42. Schulze PSC, Brindley C, Fernández JM, Rautenberger R, Pereira H, Wijffels RH, Kiron V (2020) Flashing light does not improve photosynthetic performance and growth of green microalgae. Bioresour Technol Rep 9:100367. https://doi.org/10.1016/j.biteb.2019.100367
43. Slegers PM, Wijffels RH, Van Straten G, Van Boxtel AJB (2011) Design scenarios for flat panel photobioreactors. Appl Energy 88:3342–3353. https://doi.org/10.1016/j.apenergy.2010.12.037
44. Stramski D, Sciandra A, Claustre H (2002) Effects of temperature, nitrogen, and light limitation on the optical properties of the marine diatom *Thalassiosira pseudonana.* Limnol Oceanogr 47:392–403. https://doi.org/10.4319/lo.2002.47.2.0392
45. Takache H, Pruvost J, Marec H (2015) Investigation of light/dark cycles effects on the photosynthetic growth of chlamydomonas reinhardtii in conditions representative of photobioreactor cultivation. Algal Res 8:192–204. https://doi.org/10.1016/j.algal.2015.02.009
46. Tredici MR, Carlozzi P, Chini Zittelli G, Materassi R (1991) A vertical alveolar panel (VAP) for outdoor mass cultivation of microalgae and cyanobacteria. Bioresour Technol 38:153–159. https://doi.org/10.1016/0960-8524(91)90147-C
47. Tredici MR, Zittelli GC (1998) Efficiency of sunlight utilization: tubular versus flat photobioreactors. Biotechnol Bioeng 57:187–197. https://doi.org/10.1002/(SICI)1097-0290(19980120)57:2%3c187::AID-BIT7%3e3.0.CO;2-J
48. Vu TT, Stolyar SM, Pinchuk GE, Hill EA, Kucek LA, Brown RN, Lipton MS, Osterman A, Fredrickson JK, Konopka AE, Beliaev AS, Reed JL (2012) Genome-scale modeling of light-driven reductant partitioning and carbon fluxes in diazotrophic unicellular cyanobacterium

cyanothece sp. ATCC 51142. PLoS Comput Biol 8:e1002460. https://doi.org/10.1371/journal.pcbi.1002460
49. Vunjak-Novakovic G, Kim Y, Wu X, Berzin I, Merchuk JC (2005) Air-lift bioreactors for algal growth on flue gas: mathematical modeling and pilot-plant studies. Ind Eng Chem Res 44:6154–6163. https://doi.org/10.1021/ie049099z
50. Wu X, Merchuk JC (2004) Simulation of algae growth in a bench scale internal loop airlift reactor. Chem Eng Sci 59:2899–2912. https://doi.org/10.1016/j.ces.2004.02.019
51. Wu X, Merchuk JC (2001) A model integrating fluid dynamics in photosynthesis and photoinhibition processes. Chem Eng Sci 56:3527–3538. https://doi.org/10.1016/S0009-2509(01)00048-3
52. Zhang S, Kim T-H, Han TH, Hwang S-J (2015) Influence of light conditions of a mixture of red and blue light sources on nitrogen and phosphorus removal in advanced wastewater treatment using Scenedesmus dimorphus. Biotechnol Bioproc E 20:760–765. https://doi.org/10.1007/s12257-015-0066-4

The Use of Microalgae as Bioremediation in Wastewater and Bioelectricity Production: A Review

Muhamad Maulana Azimatun Nur, Raden Herjun Desta Kresna Harsono, Amanda Nur Affrida, Fini Widya Lestari, Tia Ambar Sari, Shinta Justicia Dellarohita, Salsabila Fara Syakira, and M. Hasanuzzaman

Abstract Microalgae are microscopic photosynthetic organisms distinguished by their capacity to store carbon and develop quickly, and present a hopeful opportunity for producing sustainable biofuels and addressing environmental issues. This research investigates combining microalgae growth with biophotovoltaic (BPV) systems to simultaneously trap carbon dioxide and generate power. Through optimizing parameters such as cultivation conditions, electrode materials, and system configuration, we show notable improvements in power generation and carbon storage rates. Additionally, microalgae's ability to effectively remove contaminants and nutrients from wastewater highlights its potential for environmentally friendly wastewater treatment. The results of this study add to the expanding information on microalgae-related technologies, highlighting their importance in tackling worldwide issues like climate change and limited resources. Possible future research areas involve expanding production capacity, creating resilient strain modification techniques, and investigating new uses for microalgae biomass.

Keywords Bioremediation · Biophotovoltaic · Carbon capture · Microalgae · Wastewater

M. M. A. Nur (✉) · R. H. D. K. Harsono · A. N. Affrida · F. W. Lestari · T. A. Sari · S. J. Dellarohita · S. F. Syakira
Department of Chemical Engineering, Faculty of Industrial Engineering, Universitas Pembangunan Nasional Veteran Yogyakarta, Jl Padjajaran 104, Sleman 55283, Indonesia
e-mail: lanaazim@upnyk.ac.id

M. Hasanuzzaman
Higher Institution Centre of Excellence (HICoE), UM Power Energy Dedicated Advanced Centre (UMPEDAC), University of Malaya, Kuala Lumpur, Malaysia

A. Khan and K. A. Alzahrani (eds.), *Microalgae Biotechnology and Sustainable Development*, Sustainable Materials and Technology,
https://doi.org/10.1007/978-981-95-2597-3_2

1 Introduction

The demand for fossil fuels increases with population and economic growth. More fuel use will result in more carbon dioxide release into the atmosphere due to combustion reactions. Carbon dioxide accounts for more than 60% of greenhouse gases that can cause global warming and climate change in the world [1]. Therefore, efforts need to be made to reduce carbon emissions by developing carbon capture technologies that utilize environmentally friendly renewable energy sources. One innovative approach that can be used as a carbon capture method is to use biomass from microalgae [2]. Microalgae are single-celled photosynthetic microorganisms or colonies that can be found in a variety of humid aquatic environments [3]. Microalgae have become a potential substrate for the production of biofuels such as biodiesel, biogas, biomethane, bioethanol, and biohydrogen and can be utilized as additives in the cosmetic, health, and agricultural industries. In addition, microalgae also have several advantages, namely fast biomass growth rate, the ability to capture 10–50 times more CO_2 capture than other plants, does not require large land area, and has the ability to grow in freshwater, saltwater, and wastewater [4].

Microalgae are able to fix carbon through photosynthetic reactions, where microalgae photosynthetic carbon can contribute half of the total photosynthetic carb fixation on earth [5]. Microalgae can also be used for biological wastewater treatment as they are able to absorb pollutants in wastewater. Palm oil mill effluent (POME) provides nutrients rich in nitrogen and phosphorus, which can support microalgae growth. In addition, the cell structure of microalgae has a large surface area, allowing for rapid growth and cell division. Hariz et al. [6]. These characteristics allow microalgae to consume nutrients from POME quickly so that it can be used to manage flue gas by CO_2 capture [7]. Microalgae as a biological management agent is considered to be more environmentally friendly than the use of chemicals. However, there are still some challenges due to algal blooms that occur as a result of excess nutrients in the system. This can lead to decreased efficiency in nutrient uptake, affecting pollutant capture performance [7]. The problem can be prevented by an optimized microalgae management system considering light intensity, mixing composition, cultivation period, and nutrient concentration.

Biophotovoltaic cell (BPV) is an electrochemical device to produce electrical energy derived from the activity of photosynthetic microorganisms by converting light into chemical energy [8]. BPV consists of a cathode and anode chamber separated by a proton exchange membrane that operates through redox reactions. The reaction of these microorganisms depends on the presence of light and water. After absorbing light, water molecules break apart and release electrons, protons, and oxygen. The electrons are transferred to the catalytic cathode where they can react with protons and oxygen Qi et al. [9]. Some electrons will be used to synthesize Nicotinamide Adenine Dinucleotide Phosphate (NADPH) and Adenosine Triphosphate (ATP). Other electrons will be released and move toward the electrodes heterogeneously, then travel from the anode to the cathode to induce current flow and produce an open-circuit voltage. Excess electrons are removed through the electron transport pathway to prevent cell damage [8]. Electron transport occurs in the thylakoid membrane, cytoplasmic membrane, and outer membrane. Electron transmission from the microorganism to the electrode is provided by the biofilm. After the whole mechanism takes place, the microorganisms will utilize the materials in this process to convert carbon dioxide and organic matter into carbohydrates and proteins [10]. Thus, BPV is assumed to generate electricity in the dark by utilizing the oxidation of intracellular metabolites of microorganisms. In addition to generating bioelectricity, BPV can also sequester carbon, a means of bioremediation, and also as biomass production [11].

From the analysis of the benefits of microalgae as carbon capture and electricity generation, this study discusses the mechanism of action of microalgae in generating electricity through biophotovoltaic systems and the concept of integration of carbon capture systems. In addition, there is also a discussion of the future potential of microalgae used in liquid wastewater management and the potential for immobilization of microalgae cells for power generation.

2 Biophotovoltaic System

Biophotovoltaic (BPV) is a modern technology that combines the advantages of the electrochemical system and photovoltaic in microoganisms to produce electricity by capturing light [12]. BPV system utilizes the presence of eugenic photosynthetic microbes such as green algae and cyanobacteria to form biofilm in an electrode so that it can break water into protons, electrons, and oxygen [13]. The electrons are then transmitted to the catalytic cathode, where oxygen reacts with electrons and protons to produce water. The transport and generation of electrons occur at the microscopic level in the cytoplasmic, thylakoid, and outer membranes.

In the Biophotovoltaic circuit, the setting of the device certainly affects the power density issued by BPV, which includes a layer of biofilm and electrode material [12]. The performance of the BPV system can certainly be increased by optimizing electrode configurations and modifying electrodes. In addition, performance can also be increased by adjusting the type of device and the size of the tool, and the power density released can also be increased by controlling the external circuit operating mode.

Electricity production from microalgae in bio-photovoltaic (BPV) systems is influenced by several factors, including the choice of microalgae species, electrode material, biofilm formation, light intensity, nutrient availability, dissolved oxygen levels, and environmental conditions like pH and temperature [14]. Species with high photosynthetic efficiency, such as *Chlorella vulgaris* and *Synechocystis* sp., typically yield higher power outputs. The electrode material plays a critical role in electron transfer, with carbon-based electrodes being preferred for their high conductivity and biocompatibility. A well-established microalgae biofilm on the anode surface enhances electron transfer, though excessive biofilm thickness can reduce efficiency due to mass transfer limitations. Light intensity must be carefully balanced, as insufficient light limits photosynthesis while excessive light may cause photoinhibition [15]. Nutrient availability, particularly nitrogen and phosphorus, directly affects metabolic activity and electron generation, while oxygen produced during photosynthesis can inhibit power output by competing for electrons [16], 17]. Additionally, optimal pH (neutral to slightly alkaline) and moderate temperature conditions are crucial for maintaining metabolic rates and enhancing electricity production. Proper optimization of these factors is essential for maximizing the efficiency of BPV systems. However, during the cultivation of microalgae in wastewater, the two main factors (biofilm formation and electrode materials) should be considered.

2.1 *Biofilm Formation*

Microalgae species certainly affect the power density issued by the BPV system. Until now, species used as a trial of the BPV system are green algae, cyanobacteria, and mixed culture, according to research conducted by Qi et al. [12]. Previous study demonstrated the production of electricity from green algae including *Chlorella vulgaris, Chlorella umacc 313, Chlorella vulgaris NIES-642, Chlamydomonas reinhardtii, Paulschulzia pseudovolvox,* and *Choricystis,* as well as cyanobacteria such as *Arthrospira maxima, Pseudanabaena limnetica, Anabaena variabilis, Spirulina platensis, Synechococcus elongatus, Synechococcus bacillaris, Synechocystis sp. PCC6803, Synechococcus sp. WH5701, Nostoc sp. ATCC 27,893, and Synechococcus sp. PCC7942.* Among these, *Chlorella vulgaris and Synechocystis sp. PCC6803* exhibited the highest power outputs, generating 1900 mW/m^2 and 1620 mW/m^2, respectively. In contrast, bio-photovoltaic (BPV) systems utilizing mixed cultures did not show significant power output improvements compared to pure cultures of green algae and cyanobacteria [18].

2.2 *Electrode Material*

The selection of electrode materials affects the density and output of the power produced by BPV, where according to Qi et al. [12], electrode material affects about 20% of all BPV studies that have been conducted. The properties of anode materials such as the wisdom of the surface and the characteristics of electrochemicals have a significant impact on the power output of the BPV system. According to research conducted by [18] comparing four anode materials for BPV power generation, namely Indium Tin Oxide (ITO), Carbon Paper (CP), Rust-resistant steel (SS), and Polianilin (Pani), the surface roughness of the anode is linked to CP > PAI > SS > Ito, while the nature of the optical response shows the opposite sequence. From this, it produces data that the maximum output density of the BPV system achieved is the ITO anode material. Then Ng et al. [19] analyze the ITO anode to find out the generation of electricity out of 4 microorganisms, which shows the best performance results are *Synechococcus elongatus (UMACC 105)* > *Chlorella vulgaris (UMACC 051)* > *Chlorella* sp. *(Umacc 313)* > *Spirulina platensis.*

Electrode arrangement, as well as the qualities of the electrode material contact, has a substantial impact on BPV power production. Anode configuration evolved throughout four generations of BPV advancements. Recently, according to studies conducted by Ng et al. [19] the electrode of 3D print hierarchical pillars was analyzed to generate electricity significantly by overcoming electrolyte deficiency and light penetration. The researchers threw the ITO nanoparticles into the electrode plate in order to form a micropillar-appraisal electrode. In this case, the electrodes of the pillar of branches and the electrical density of BPV output reached 2450 mA/m^2 power; this number approached the theoretical predictions studied.

3 Principle of Photosynthetic Carbon Sequestration in Microalgae

Light and dark reactions are the main reactions in photosynthetic carbon uptake in microalgae [20, 21]. According to Fig. 2, the photosynthetic process of microalgae includes an electron transfer process, in which electrons move from water molecules to two photosynthetic systems, which are Photosystem I (PSI) and Photosystem II (PSII) [22]. The PSII complex initiates the process by transferring electrons to primary acceptors with the help of photons and chlorophyll. The electrons proceed through an electron transport system along the membrane, where PSI obtains electrons from PSII and carries them through chlorophyll. In the end, photosynthesis generates NADPH and ATP through light-driven reactions, allowing ATP to store energy via photophosphorylation for use in the dark reaction process [23] (Fig. 1).

The process of the dark reaction involves three phases. In the first stage, during carboxylation, ribulose-1,5-bisphosphate carboxylase assists the reaction between CO_2 and ribulose-1,5-bisphosphate, which produces two molecules of 3-phosphoglycerate (3-PGA). Next, during the reduction process, ATP converts 3-PGA into 1,3-diphosphoglycerate via 3-phosphoglycerate kinase, followed by NADPH converting it into glyceraldehyde-3-phosphate (G-3-P) with the help of glyceraldehyde phosphate dehydrogenase. Ultimately, during regeneration, G-3-P is converted back to ribulose-1,5-bisphosphate using ATP and enzymes, leading to the capture of CO_2 and the manufacture of photosynthetic substances such as starch and glucose. Research on biomass production in microalgae Dunaliella tertiolecta SAD-13.86 has demonstrated that CO_2 fixation plays a crucial role in the production of biomass, which consists predominantly of proteins and also contains a significant amount of lipids [24]. The increasing understanding of CO_2 fixation pathways indicates a bright future for microalgae in sustainable biomass and product manufacturing.

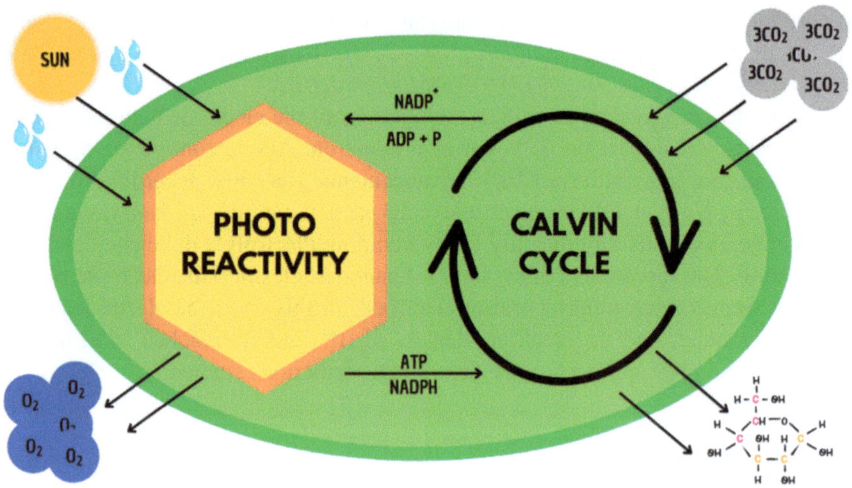

Fig. 1 Schematic diagram of microalgae photosynthetic carbon fixation

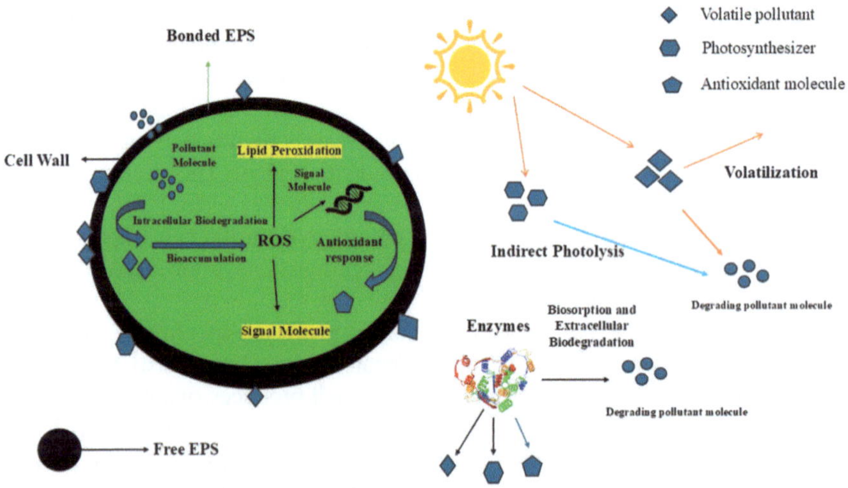

Fig. 2 Principle of pollutant reduction by microalgae [28]

3.1 *Principle of Pollutant Reduction by Microalgae*

Biological treatment of wastewater is being improved with advances in pollutant detection, stricter environmental regulations, and improved treatment technologies [25]. One ingenious and environmentally benign method in biological treatment is the use of microalgae for the degradation of organic and inorganic impurities. Microalgae

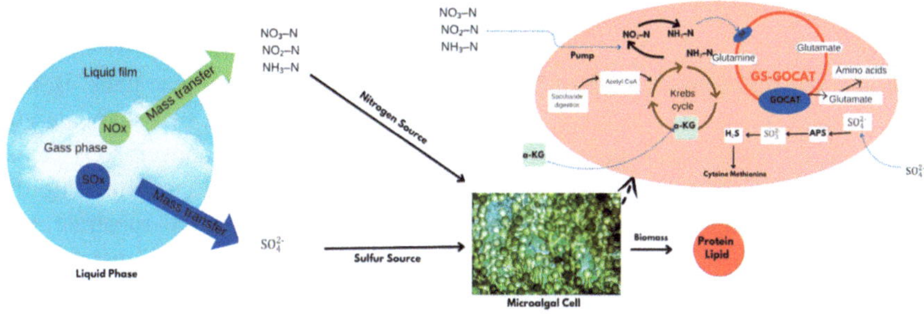

Fig. 3 Schematic diagram of NOx/SOx emission reduction by microalgae [28]

perform waste degradation through three mechanisms: biosorption, bioconcentration, and biodegradation [26, 27]. As illustrated in Fig. 2, microalgae cells reduce pollutants through these mechanisms.

When functional groups on pollutant molecules interact with charged groups on the surface of microalgal cells, biosorption is triggered, resulting in the binding of pollutants to the cell surface through electrostatic interactions and chemical modifications [28]. According to [29], the extracellular polymers that envelop the microalgal cells, such as polysaccharides and soluble and insoluble proteins, promote biosorption. Pollutants enter microalgal cells by active transport, facilitated diffusion, or passive diffusion, whereupon they bioaccumulate. Furthermore, microalgae can biodegrade organic contaminants into non-toxic or less damaging molecules while using the organic matter as a source of carbon for growth and development [30]. Biodegradation helps to remove pollutants from microalgae cells and decreases oxidative stress.

In addition to direct mechanisms, microalgae can also remove pollutants through indirect processes such as photodegradation and volatilization, as depicted in Fig. 3. [31] investigated the photodegradation of aniline using freshwater microalgae, and found that the removal of aniline was due to reactive oxygen species in microalgae rather than biosorption or bioaccumulation. Similarly, microalgae have shown effective removal of new pollutants, such as emerging pollutants [32]. Various studies have demonstrated the ability of microalgae to degrade emerging pollutants such as hormones, antibiotics, pesticides, and analgesics, making microalgal systems a safe choice for wastewater pollutant treatment.

3.2 *Sulfur and Nitrogen Oxides Reduction by Microalgae*

Air pollutants SOx and NOx serve as nitrogen and sulfur sources for the growth of some microalgae species, which can be used to make biofuels, animal protein feed, and natural fertilizers [33]. NO and NO_2, two common forms of NOx, react in the presence of O_2 in water to form HNO_2 and HNO_3 according to reactions (1–4) [34,

35), specifically, $2NO_2 + O_2 = 2NO_2$ (1), $4NO + 2H_2O + 3O_2 = 4HNO_3$ (2), $2NO_2 + H_2O = HNO_3 + HNO_2$ (3), and $4NO_2 + 2H_2O + O_2 = 4HNO_3$ (4). Figure 4 describes the main mechanism of NOx synthesization by microalgae cells. Before these compounds are assimilated, microalgae need to undergo a two-step catalytic reduction process using nitrate reductase and nitrite reductase to convert NO_3-N and NO_3-N into NH_3-N. NH_1-N, NO_1-N, and NO_2-N are actively transported into the cell at the microalgal plasma membrane. Acetyl-CoA is the initial substrate, and α-ketoglutarate (α-KG) from the Krebs cycle is involved in microalgal metabolism. Glutamate synthetase (GS) and glutamine ketoglutarate aminotransferase (GOGAT) regulate the GS-GOGAT pathway. Glutamic acid is created when α-KG and NH_3-N combine, and it is then utilized to synthesize amino acids. Glutamate synthetase, which catalyzes the conversion of glutamate to glutamine using ATP, directly converts NH_3-N, which results from the reduction of NO_3-N and NO_2-N, into amino acids [36].

Ma et al. [33] isolated a functional microalgae strain PF3 capable of removing NO and demonstrated its use for the biological removal of NO and CO_2 in simulated flue gas. Microalgae have a preference for the most reduced form of nitrogen, with the order of utilization being $NH_4 + > NO > NO_2^- > NO_3^- > CO(NH_2)_2$ [33]. Sulfur is necessary for SOx because it helps microalgae make sulfur-containing thylakoid lipids and important amino acids like cysteine and methionine [37]. The schematic mechanism of microalgae cells' SOx utilization is depicted in Fig. 4. Sulfate is taken up by microalgae into the cytoplasm and transported by a high-affinity sulfate transport system to plastids or vacuoles. Adenosine-5′-phosphosulfate (APS), which is subsequently reduced to $SO_3 2-$ by APS reductase and further reduced to sulfide

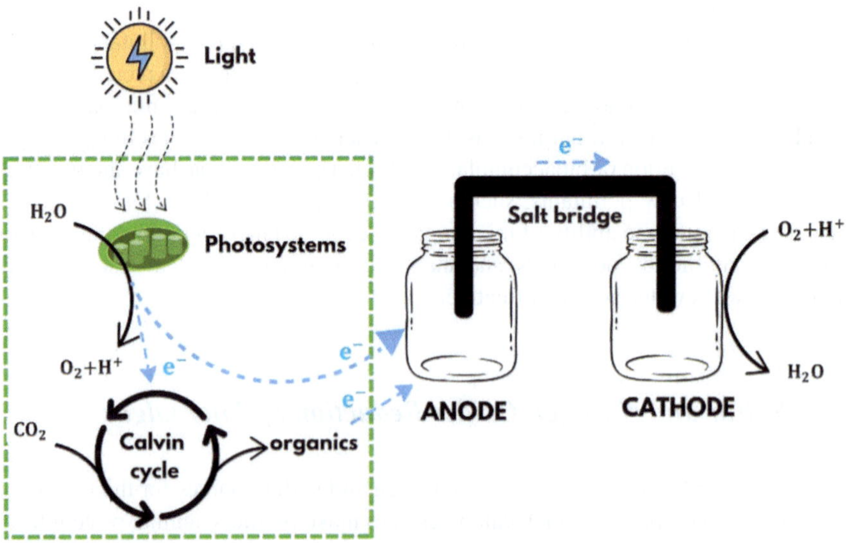

Fig. 4 A brief schematic representation of BPV systems

by ATP sulfurylase, is created when ATP activates this sulfate ion [38]. Du et al. [39] used microalgae to successfully combine the production of lipids with the removal of SOx.

Figure 3 illustrates the integration of nitrogen and sulfur metabolism in microalgae to produce biomass, proteins, and lipids. Nitrogen oxides (NOx) and sulfur oxides (SOx) are absorbed from the gas phase into the liquid layer, where they undergo mass transfer into the liquid phase. Nitrogen sources such as NO_2-N, NO_3-N, and NH_4-N, and sulfur sources such as SO_4^{2-} are taken up by microalgal cells. Within microalgal cells, these compounds participate in important metabolic processes, including the nitrogen cycle, in which NO_2-N and NO_3-N are reduced and integrated into amino acids, and the sulfur cycle, which involves the conversion of sulfate into organic sulfur compounds.

3.3 Microalgae in Producing Bioelectricity in BPVs

BPV involves the process of oxygen photosynthesis performed by photosynthetic microorganisms [40, 41]. Phototrophic organisms are divided into two categories: oxygenic and anoxygenic phototrophs. Oxygenic phototrophs only participate in oxygen photosynthesis and use water as an electron donor. This process is facilitated by the photosynthetic machinery, which includes the collaborative functions of photosystem I (PSI) and photosystem II (PSII) [42]. In oxygenic photosynthesis, an energy transfer process occurs that involves electron transfer, electron storage, light charge separation, and light absorption. Initially, PSII generates photosynthetic electrons through the photolysis of water. These electrons are subsequently relayed through the photosynthetic electron transport chain (PETC), involving components such as plastoquinones (PQ/PQH2), the cytochrome b6f complex (cyt bf), plastocyanin (PC), PSI, and ferredoxin (Fd), ultimately leading to the conversion of electrons into NADPH by ferredoxin-NADP + reductase (FNR) for the purposes of carbon fixation and cellular growth. To convert light energy into electricity in BPV systems, an additional mechanism is required to transfer electrons from inside the cell to the electrode outside the cell, called exoelectrogenesis or extracellular electron transfer (EET). In theory, electrons from the photosynthetic process can be taken directly from excited PSII/PSI or from reduced energy carriers such as PQH2, NADPH, and carbon fixation products through oxidation reactions. After the photosynthetic electrons are released from the microorganisms, the BPV process continues in the bioelectrochemical system, where the electrons are first captured by the anode. At the same time, high-potential reactions such as oxygen reduction occur at the cathode, creating a potential difference between the anode and cathode that promotes the flow of electrons through an external circuit, thereby generating a light-dependent electric current.

Figure 4 illustrates the integration of photosynthesis with a bio electrochemical system (BES), demonstrating the potential of light energy to drive electron transfer for organic production and energy generation. In the photosystems of a photosynthetic organism, light energy facilitates the splitting of water (H_2O) molecules into oxygen (O_2), protons (H^+), and electrons (e^-). The Calvin cycle then uses CO_2 and the products of photosynthesis to synthesize organic compounds. Meanwhile, the generated electrons are transferred through an external circuit to the anode of the BES, then flow to the cathode via a salt bridge, completing the circuit. At the cathode, oxygen reduction occurs, producing water. This setup showcases the synergy between biological photosynthetic processes and electrochemical systems, which can be harnessed for sustainable bioenergy production and carbon fixation into valuable organic compounds.

As shown in Table 1, the highest microalgae growth occurs on day 12. The concentration obtained is the highest concentration in the batch method which is around 0.55 g/L. Based on the literature, an increase in power density will correlate with algae concentration; this is because algae are supplied by O_2 [47]. The increase in dissolved oxygen content at the cathode at the start of the test is proportionate to the strong photosynthetic activity of algae during the growth phase [44]. It can be concluded that the oxygen produced by the algae during photosynthesis not only affects the growth of the algae themselves, but also the electrochemistry around the cathode.

The cathode voltage is largely determined by how much dissolved oxygen is present, where oxygen molecules are used as electron receivers in reactions that occur at the cathode. However, the accumulation of biological material and salt deposits that impede the flow of substances that receive electrons might impair the cathode's functionality [49]. As a result, the electron mass transfer rate is limited, which will impact the efficiency of the system. Furthermore, according to a different study, the redox reaction that results from oxygen migration to the anode across the cation exchange membrane may be positive [50]. A substantial oxygen concentration difference between the anode and cathode compartments and the presence of catalysts on the anode surface that speed up the oxygen reduction reaction are two possible causes of this process. Microorganisms at the anode oxidize organic substrates, releasing electrons that are used to create power [51]. High substrate concentrations have been shown in the literature to drastically reduce the activity of exoelectrogen bacteria by reducing their ability to function [52]. These conditions cause the anode to have a relatively high potential. This high potential indicates that fewer electrons are being transferred to the anode, reducing the efficiency of electricity production. This suggests that there is an optimal substrate concentration that maximizes electricity production.

Yang et al. [44] suggested that algae can efficiently absorb nitrogen and phosphorus for biomass synthesis. In line with that, according to Research [53] studied the removal of organic matter, nitrogen, and phosphorus carried out in batch and continuous operation modes. During continuous operation, the phosphorus removal efficiency in the catholyte ranged from 78.1 to 91.1%, which is much higher than that in batch operation mode. Algae have the potential to accumulate large amounts of phosphorus removal through absorption. In addition, the continuous operation mode

Table 1 Electrical performance of microbial and algal BPV units using various wastewaters

No	Strains	Wastewater utilized	Maximum power density	References
1	*Shewanella* sp. *Enterobacter* sp. *Bacillus* sp. *Clostridium* sp.	Municipal solid waste leaching water	95.63 mW/m^2	Lakshmidevi et al. [43]
2	Members of Clostridia and Betaproteobacteria (most dominant)	Anaerobic digestion of kitchen waste	2.34 W/m^3	Yang et al. [44]
3	*Coelastrella* sp.	Wastewater from a dairy products processing plant	31.4 mW/m^2	Raja et al. [45]
4	*Pseudomonas* *Methylopila* *Methylobacterium* *Clostridium*	Anaerobic digestion of kitchen waste	3.5 W/m^3	Yang et al. [44]
5	*Leptolyngbya* sp.	Raw wastewater	0.008 mW/cm^2	Maity et al. [46]
6	*Chlorella* sp.	Palm oil mill effluent (POME)	0.45 mW/m^2	Ng et al. [77]
7	*Terrimonas ferruginea, Bradyrhizobium Alphaprot eobacteria*	Synthetic wastewater	68 mW/m^2	Zhang et al. [47]
8	*Chlamydomonas sp.*	Wastewater remediation	1.83 mW/m^2	Behl et al. [48]

has a nitrogen removal efficiency of 22%, which is higher than that in batch operation mode. Furthermore, in continuous operation mode, the ammonium removal efficiency from the anolyte reaches 93.5%.

Yang et al. [44] proposed that algae can efficiently absorb nitrogen and phosphorus for biomass synthesis. In keeping with that, according to research, [53] investigated the removal of organic matter, nitrogen, and phosphorus using batch and continuous operating modes. During continuous operation, the catholyte's phosphorus removal effectiveness ranged from 78.1 to 91.1%, which was much greater than in batch operation. Algae have the capacity to collect significant levels of phosphorus elimination via absorption. Furthermore, the continuous operating mode achieves a nitrogen removal effectiveness of 22%, which is higher than the batch operation mode. Furthermore, in continuous operating mode, the ammonium removal efficiency from the anolyte reaches 93.5 percent. Previous studies have shown that the loss of ammonium in a two-chamber microbial fuel cell (MFC) is mainly due to the movement of ammonium from the anode chamber to the cathode chamber through the cation exchange membrane (CEM). The increasing collection of ammonium in the cathode chamber in batch mode confirms that most of the ammonium in the anode chamber moves to the cathode chamber. This ammonium movement is mainly caused by two things: the difference in the amount of ammonium between the two chambers and the effort to maintain charge balance. It is estimated that the ammonium migration rate is about 2.0 g per square meter per day in continuous mode and 1.7 g per square meter per day in batch mode. This rate is lower than the results of other studies using higher ammonium concentrations.

At the beginning of the production cycle, the amount of ammonium in the catholyte is relatively low. However, the amount increases significantly at the end of the cycle. This is likely because some ammonium evaporates due to the slightly alkaline pH conditions in the catholyte. In addition to ammonium, small amounts of other substances such as nitrite and nitrate are also found in the catholyte. The amount of these substances decreases when the dissolved oxygen (DO) level decreases. However, when ammonium is very high, algae prefer to use ammonium rather than nitrate or nitrite as a nutrient source. Nitrate and nitrite can be converted into other forms at the cathode. During the continuous production process, the amount of ammonium decreases, while the amount of nitrate and nitrite increases. Ammonium decreases because it is used by algae or converted into nitrate and nitrite by oxygen produced by algae through photosynthesis. Meanwhile, nitrate and nitrite can also be converted into other forms at the cathode [44]. However, the process of reducing the amount of nitrate and nitrite under continuous production conditions is hampered by high dissolved oxygen levels. High oxygen levels actually cause an increase in the amount of nitrate and nitrite, and inhibit the process of changing nitrate and nitrite into other forms at the cathode [47].

4 The Future Potential of Microalgae for Carbon Capture and Electricity Generation in Wastewater Cultivation

For the industrial sector, microalgae's ability to successfully sequester carbon and manage waste is essential. It not only opens up economic opportunities with great potential, but also provides solutions to global environmental challenges. In addition to eliminating contaminants from the waste, the use of microalgae in waste treatment can also indirectly lower expenses [54]. Microalgae may generally be utilized to cleanse water more effectively, preserving water quality and lessening the effects of environmental harm. Compared to traditional technologies, carbon fixation technology by microalgae has several advantages, namely that it can convert CO_2 into biomass and can produce products that have high value [36, 55]. Currently, the utilization of microalgae for carbon fixation and wastewater treatment provides a very promising opportunity if further developed [56].

Microalgae present significant opportunities for future applications in various industrial sectors due to their substantial potential. Microalgae, being a renewable and eco-friendly resource, present sustainable solutions by producing bioelectricity and capturing carbon dioxide [57]. Furthermore, high-valueable compounds from the biomass with a range of industrial uses, including pigments, proteins, and fatty acids, with various industrial applications [58]. Nevertheless, further research is required to fully realize their industrial-scale potential. Since microalgae effectively remove excess nutrients and organic contaminants from wastewater, they help improve wastewater quality and support sustainable development, which has a significant positive impact on the environment.

With this much potential, technology that accommodates it is also needed. As is known that microalgae based on research that has been done depends on the intensity of sunlight, therefore, if it is not supported by power storage that accommodates it, it will be very useless because microalgae will only fulfill electricity around 12–27% of the electricity needed [59]. Thus, technological advancements in energy distribution and storage must go hand in hand with the growth of renewable energy. Storage is necessary after microalgae biomass is converted into solid liquid fuel so that it can be utilized whenever needed. Preservation is the primary goal of storage, not extraction or purification [60].

Using wastewater as the substrate, anaerobic bacteria in the anode of a microbial fuel cell (MFC) in Fig. 5 undergo anaerobic digestion to release protons (H +) and electrons (e-). While the protons move through a membrane to the cathode, the electrons travel through an external circuit to produce electricity. In the cathode, microalgae absorb the CO2 released from the anode and, through photosynthesis, contribute to the reduction process, utilizing the electrons and protons to complete the circuit. The design showcases the dual function of bioelectricity generation and CO_2 fixation, with light supporting photosynthetic activity in the cathode. Previous research demonstrated that the integrated process using *Enterobacter aerogenes* with sodium alginate and *Nostoc* sp. as the cathode resulted in 83.15% COD removal and a power density of 168 mW/m^2 [61]. He et al. [76] also found that *Chlorella vulgaris* immobilized with 5% sodium alginate as the cathode, combined with bacterial sludge in the anode, achieved a significantly higher power density of 2572.8 mW/m^3. Similarly, [62] used *C. vulgaris* immobilized in a 4% sodium alginate solution as the cathode, with bacterial sludge in the anode, producing a power density of 2485.35 mW/m^3. These research demonstrate how bacteria-microalgae systems can be used to produce important products from microalgae, such as carbon capture, wastewater treatment, and energy.

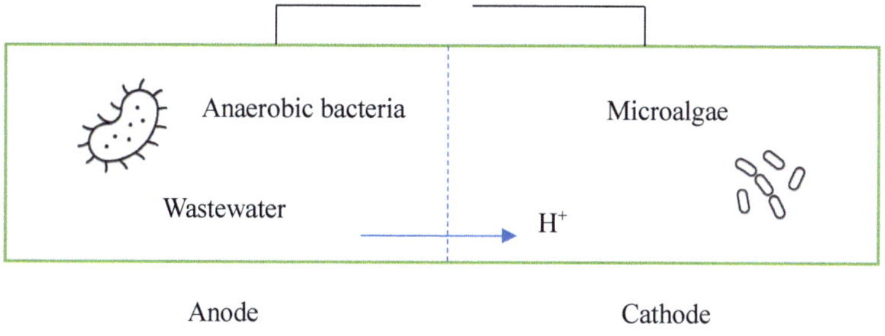

Fig. 5 Integration of bacteria-microalgae producing electricity, wastewater treatment, and carbon capture removal

5 Potential of Microalgae Cell Immobilization for Electricity Generation

In recent research, efforts to enhance the output power generated by algal-based BioPhotovoltaic (BPV) systems have been extensive. One approach involves the immobilization of microalgae. This process is considered to have the potential to increase the electrical power produced by BPVs. In microalgae cultivation, immobilization techniques are methods that restrict the movement of algal cells in aqueous media by trapping them within a solid matrix [63]. Gel entrapment is the most common method for immobilizing microalgae cells. Agar, carrageenan, and alginate are examples of natural polysaccharides that are good for immobilizing microalgae because of their high transparency and low toxicity [64].

Figure 6 demonstrated the immobilization of microalgae to produce electricity. Immobilized cell microalgae produce bio-based products (BPV) through a process where microalgae cells are physically confined or attached to a support matrix while maintaining their metabolic activity. In this setup, the microalgae can continuously interact with their environment, such as light and nutrients, facilitating efficient biomass accumulation and product formation [65].

The main advantage of immobilization is the ease with which the cells can be easily harvested when growing on a wastewater. The immobilization supports improved cell stability and retention, enhancing the overall productivity and ease of product recovery. As the microalgae grow, they undergo photosynthesis, converting carbon dioxide and light into valuable products, including electricity generation. The immobilized system often allows for repeated use of the microalgae, which can be advantageous for scaling up production processes and reducing operational costs. This approach leverages the high productivity of microalgae while addressing challenges related to cell separation and process efficiency (Fig. 7).

The most common natural polysaccharide utilized in algal immobilization research is alginate. It is usually generated commercially as sodium or calcium alginate and is formed from the cell walls of macroalgae found in the brown sea. The output power is determined by the amount of electrons transferred through this alginate. Within the gel matrix, individual algae cells in immobilized cell systems begin

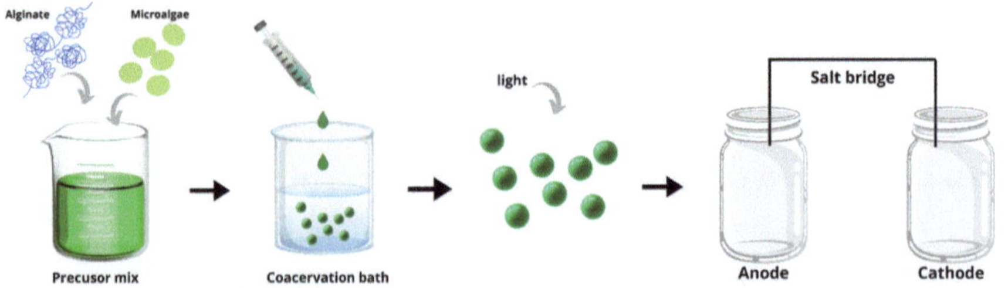

Fig. 6 A brief schematic representation of algae immobilization entrapment onto a matrix and its application onto BPV systems

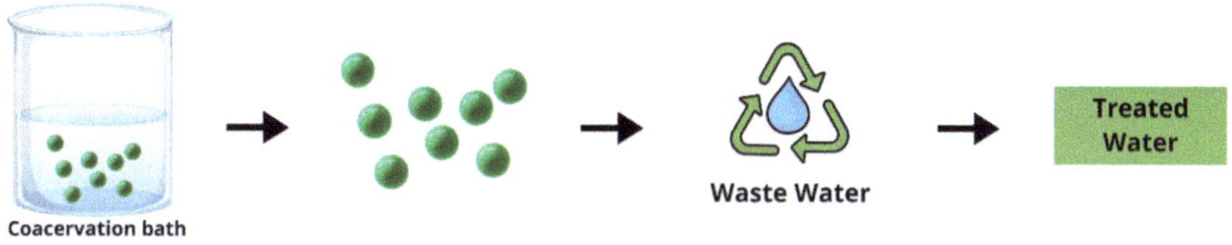

Fig. 7 Schematic process of wastewater pollutant reduction by immobilized microalgae [65]

to divide and form colonies. This differs from suspension cultures where cells can be dispersed in aqueous media. Alginate helps to lock or immobilize the algal cells, keeping them in a fixed position and preventing them from moving freely in the liquid medium. As a result, it is easier to position the algae cells in closer proximity to the anode, increasing the effectiveness of electron transmission from the source to the anode [66]. The colonial character of alginate, which efficiently links algae cells inside the immobilized system and facilitates quicker transfer, is another reason it was selected. The performance of immobilized cell photosynthesis confirms suspension culture methods in terms of relative electron transfer rate measure (rETRm), but immobilized cultures form colonies that can improve cellular connectivity and increase electron transport in BPVs. Longer periods of photosynthetic efficiency can be sustained by immobilized cells.

In previous research, Indium Tin Oxide (ITO) anodes was used in this immobilization process as they are the most suitable and efficient for microalgae immobilization. Furthermore, the development of algal biofilms can profit greatly from ITO's optical transparency. The rate of electron transmission to the electrode margins is still a major obstacle, though [19]. Reduced Graphene Oxide (RGO), an alternative anode with inherent electrical qualities and biological compatibility, may increase efficiency. Moreover, the Langmuir–Blodgett method has proven to be an alternative that can increase power output by up to 119% [19].

Table 2 presented a summary of research that examined the use of various matrices, immobilizers, and microorganisms in microbial fuel cells (MFCs) to produce electricity. Several strains of microalgae and bacteria, including *Chlorella* sp., *Saccharomyces cerevisiae, Anabaena variabilis*, *Enterobacter aerogenes*, and *Chlorella vulgaris*, were immobilized using materials such as sodium alginate, polyion complexes, and alginate beads. These immobilizers support the microorganisms within MFC systems, allowing them to generate power under various conditions.

Notably, immobilized *Chlorella* sp. achieved relatively low power densities of 0.073 mW/m^2 respectively, with sodium alginate powder and carbon paper as immobilizers. In contrast, immobilized *Saccharomyces cerevisiae* generated a higher power density of 13.94 mW/m^2, highlighting its more efficient power generation capabilities compared to the microalgae. Additionally, *Enterobacter aerogenes* achieved a power density of 168 mW/m^2 when paired with *Nostoc* sp. as a cathode and exhibited significant chemical oxygen demand (COD) removal of 83.15%. Interestingly, studies involving immobilized *Chlorella vulgaris* showed substantially higher power densities (2,485.35–2,572.8 mW/m^3) when used as a cathode in systems involving

Table 2 Power output performance of immobilized algae BPV devices. NA means not applicable or not recorded

No	Strains	Matrix Immobilizer	Remarks	Maximum power density	Source
1	*Chlorella* Sp. (UMACC 313)	Sodium Alginate powder	Indium Tin Oxide (ITO)	0.073 ± 0.007 mW m^{-2}	Ng et al. [66]
2	*Saccharomyces cerevisiae*	Sodium Alginate powder	Microalgae Microbial Fuel System	13.94 mW·m^{-2}	Hadiyanto et al. [67]
3	*Synechocystis sp.* PCC6714	Poly-ion complex	Glassy carbon electrode	NA	Yagishita et al. [68]
4	*Anabaena variabilis*	Alginate beads	Discharge/Culture Cycles	0.6 mA	Yagishita et al. [69]
5	*Enterobacter aerogenes*	Sodium alginate	*Nostoc* sp. as cathode, 83.15% COD removal	168 mWm^{-2}	Lakshmidevi et al. [61]
6	*C. vulgaris*	5% sodium alginate	Immobilized *C. vulgaris* as cathode, bacteria sludge in anode	2572.8 mW/m^3	He et al. [76]
7	*C. vulgaris*	4% sodium alginate solution	Immobilized *C. vulgaris* as cathode, bacteria sludge in anode	2485.35 mW/m^3	Zhou et al. [62]

bacteria sludge at the anode. This suggests a potential synergistic effect between C. vulgaris and bacterial sludge for improved electron transfer efficiency. The choice of microorganism, matrix, and immobilizer significantly affects power generation in MFCs. Moreover, immobilization techniques and the configuration of cathodes and anodes play crucial roles in optimizing power density, indicating that tailored systems leveraging the right combinations of these variables can enhance the performance of MFCs.

6 Conclusion

Biophotovoltaic (BPV) is a modern technology that combines the advantages of the electrochemical system and photovoltaic in microorganisms to produce a electricity by capturing light. Systematically, the BPV setup technique consists of two aspects: biofilm and electrode material. The two most common microalgae species that generate the greatest power density are green algae (Chlorella vulgaris) and cyanobacteria (Synechocystis sp. PCC6803). The limits of light and electrolyte penetration can be overcome by the 3D pillar array electrode, which can greatly boost the power density.

BPV has several benefits, one of which being its ability to sequester carbon. By absorbing carbon dioxide from the atmosphere and converting it to biomass, microalgae can help mitigate the effects of climate change. In addition, BPV systems can be used to treat wastewater, remove pollutants, and improve water quality. Microalgae of Pseudomonas, Methylopila, Methylobacterium, and Clostridium species cultured in anaerobically digested kitchen waste medium showed high maximum power density values of 2.34 3.5 W m^{-3}. With this potential, microalgae can be an appropriate solution for carbon capture technology, waste treatment, and as a renewable energy source to produce electricity.

References

1. Maghzian A, Aslani A, Zahedi R (2024) A comprehensive review on effective parameters on microalgae productivity and carbon capture rate. J Environ Manag 355. https://doi.org/10.1016/j.jenvman.2024.120539
2. Zhao Y, Yan Y, Jiang Y, Cao Y, Wang Z, Li J, Yan C, Wang D, Yuan L, Zhao G (2024) Release pattern of light aromatic hydrocarbons during the biomass roasting process. Molecules 29(6). https://doi.org/10.3390/molecules29061188
3. Khanra A, Vasistha S, Rai MP, Cheah WY, Khoo KS, Chew KW, Chuah LF, Show PL (2022) Green bioprocessing and applications of microalgae-derived biopolymers as a renewable feedstock: circular bioeconomy approach. Environ Technol Innov 28. https://doi.org/10.1016/j.eti.2022.102872
4. Selim KA, Alva V (2024) PII-like signaling proteins: a new paradigm in orchestrating cellular homeostasis. Curr Opin Microbiol 79. https://doi.org/10.1016/j.mib.2024.102453
5. Deb D, Dixit A, Chandra L (2023) Smart innovation, systems and technologies.Renew Energy Clim Chang 161. http://www.springer.com/series/8767
6. Hariz HB, Takriff MS, Mohd Yasin NH, Ba-Abbad MM, Mohd Hakimi NIN (2019a) Potential of the microalgae-based integrated wastewater treatment and CO_2 fixation system to treat Palm Oil Mill Effluent (POME) by indigenous microalgae; Scenedesmus sp. and Chlorella sp. J Water Process Eng 32. https://doi.org/10.1016/j.jwpe.2019.100907
7. Hariz HB, Takriff MS, Mohd Yasin NH, Ba-Abbad MM, Mohd Hakimi NIN (2019b) Potential of the microalgae-based integrated wastewater treatment and CO_2 fixation system to treat Palm Oil Mill Effluent (POME) by indigenous microalgae; Scenedesmus sp. and Chlorella sp. J Water Process Eng 32. https://doi.org/10.1016/j.jwpe.2019.100907
8. Chin JC, Ng FL, Kang HS, Li B, Goh PS, Lim JW, Ghfar AA, Mohammed AAA, Shahid MK (2024) Effect of hydrodynamic aeration on bioelectricity generation from photoautotrophic Spirulina platensis in biophotovoltaic devices. Algal Res 77. https://doi.org/10.1016/j.algal.2023.103341
9. Qi X, Liu X, Gu Y, Liang P (2024a) Whole-cell biophotovoltaic systems for renewable energy generation: a systematic analysis of existing knowledge. Bioelectrochemistry 158. https://doi.org/10.1016/j.bioelechem.2024.108695
10. Chandrasekaran K, Nangan S, Okhawilai M, Gnanasekaran L, Salmen SH, Ali Alharbi S, Uyama H (2024) Converting waste PET plastic into sulfonated carbon nanosheets supported PANI nanocomposite as anode catalyst in biophotovoltaic cells. Fuel 362. https://doi.org/10.1016/j.fuel.2023.130583

11. Tee JY, Ng FL, Keng FSL, Lee CW, Zhang B, Lin S, Gnana kumar G, Phang SM (2024) Green synthesis of reduced graphene oxide by using tropical microalgae and its application in biophotovoltaic devices. IScience 27(4). https://doi.org/10.1016/j.isci.2024.109564
12. Qi X, Liu X, Gu Y, Liang P (2024b) Whole-cell biophotovoltaic systems for renewable energy generation: a systematic analysis of existing knowledge. Bioelectrochemistry 158. https://doi.org/10.1016/j.bioelechem.2024.108695
13. Tschörtner J, Lai B, Krömer JO (2019) Biophotovoltaics: green power generation from sunlight and water. Front Microbiol 10, Issue APR. https://doi.org/10.3389/fmicb.2019.00866
14. Zhu H, Wang H, Zhang Y, Li Y (2023) Biophotovoltaics: recent advances and perspectives. Biotechnol Adv 64:108101
15. Chandra R, Venkata Mohan S, Roberto PS, Ritmann BE, Cornejo RAS (2018) Biophotovoltaics: conversion of light energy to bioelectricity through photosynthetic microbial fuel cell technology. Microb Fuel Cell: Bioelectrochemical Syst Convert Waste Watts 373–387
16. Ng FL, Phang SM, Lan BL, Kalavally V, Thong CH, Chong KT, Fisher AC (2020) Optimised spectral effects of programmable LED arrays (PLA) s on bioelectricity generation from algal-biophotovoltaic devices. Sci Rep 10(1):16105
17. Reilly-Schott V, Gaibler J, Bai Y, Mier-Jimenez A, Qasim M, Lai B (2024). Electron leaks in biophotovoltaics: a multi-disciplinary perspective. ChemCatChem e202400639
18. Venkata Subhash G, Chandra R, Venkata Mohan S (2013) Microalgae mediated bio-electrocatalytic fuel cell facilitates bioelectricity generation through oxygenic photomixotrophic mechanism. Biores Technol 136:644–653. https://doi.org/10.1016/j.biortech.2013.02.035
19. Ng FL, Phang SM, Thong Ng F-L, Jaafar MM, Phang S-M, Chan Z, Salleh NA, Azmi SZ, Yunus K, Fisher AC, Periasamy V (2014) Reduced graphene oxide anodes for potential application in algae biophotovoltaic platforms. Sci Rep 4(1):7562. https://doi.org/10.1038/srep07562
20. Elrayies GM (2018) Microalgae: prospects for greener future buildings. Renew Sustain Energy Rev 81:1175–1191. https://doi.org/10.1016/j.rser.2017.08.032
21. Li S, Li X, Ho SH (2022) How to enhance carbon capture by evolution of microalgal photosynthesis? Sep Purif Technol 291. https://doi.org/10.1016/j.seppur.2022.120951
22. Zeng J, Wang Z, Chen G (2021) Biological characteristics of energy conversion in carbon fixation by microalgae. Renew Sustain Energy Rev 152. https://doi.org/10.1016/j.rser.2021.111661
23. Prasad R, Gupta SK, Shabnam N, Oliveira CYB, Nema AK, Ansari FA, Bux F (2021) Role of microalgae in global CO_2 sequestration: physiological mechanism, recent development, challenges, and future prospective. Sustainability (Switzerland) 13(23). https://doi.org/10.3390/su132313061
24. Sydney EB, Sturm W, de Carvalho JC, Thomaz-Soccol V, Larroche C, Pandey A, Soccol CR (2010) Potential carbon dioxide fixation by industrially important microalgae. Biores Technol 101(15):5892–5896. https://doi.org/10.1016/j.biortech.2010.02.088
25. Makhanya BN, Nyandeni N, Ndulini SF, Mthembu MS (2021) Application of green microalgae biofilms for heavy metals removal from mine effluent. Phys Chem Earth 124. https://doi.org/10.1016/j.pce.2021.103079
26. Razzak SA, Ali SAM, Hossain MM, deLasa H (2017) Biological CO_2 fixation with production of microalgae in wastewater–a review. Renew Sustain Energy Rev 76:379–390. https://doi.org/10.1016/j.rser.2017.02.038
27. Singh K, Ansari FA, Ingle KN, Gupta SK, Ahirwal J, Dhyani S, Singh S, Abhilash PC, Rawat I, Byun C, Bux F (2023) Microalgae from wastewaters to wastelands: leveraging microalgal research conducive to achieve the UN sustainable development goals. Renew Sustain Energy Rev 188. https://doi.org/10.1016/j.rser.2023.113773
28. Dahai H, Zhihong Y, Lin Q, Yuhong L, Lei T, Jiang L, Liandong Z (2024a) The application of magical microalgae in carbon sequestration and emission reduction: removal mechanisms and potential analysis. Renew Sustain Energy Rev 197. https://doi.org/10.1016/j.rser.2024.114417

29. Cuellar-Bermudez SP, Aleman-Nava GS, Chandra R, Garcia-Perez JS, Contreras-Angulo JR, Markou G, Muylaert K, Rittmann BE, Parra-Saldivar R (2017) Nutrients utilization and contaminants removal. a review of two approaches of algae and cyanobacteria in wastewater. Algal Res 24:438–449. https://doi.org/10.1016/j.algal.2016.08.018
30. Kothari R, Pathak AK, Singh HM, Goria K, Din ZU, Bharti A, Raina S, Rachna Singh A, Singh B, Kumar VA, Pandey A, Pandey AK, Pathania D, Tyagi VV (2024) MFC-mediated wastewater treatment technology and bioelectricity generation: Future perspectives with SDGs 7 & 13. Process Saf Environ Prot. https://doi.org/10.1016/j.psep.2024.08.078
31. Wang L, Zhang C, Wu F, Deng N (2007) Photodegradation of aniline in aqueous suspensions of microalgae. J Photochem Photobiol, B 87(1):49–57. https://doi.org/10.1016/j.jphotobiol.2006.12.006
32. Prosenc F, Piechocka J, Škufca D, Heath E, Griessler Bulc T, Istenič D, Buttiglieri G (2021) Microalgae-based removal of contaminants of emerging concern: mechanisms in Chlorella vulgaris and mixed algal-bacterial cultures. J Hazard Mater 418. https://doi.org/10.1016/j.jhazmat.2021.126284
33. Ma S, Li D, Yu Y, Li D, Yadav RS, Feng Y (2019) Application of a microalga, Scenedesmus obliquus PF3, for the biological removal of nitric oxide (NO) and carbon dioxide. Environ Pollut 252:344–351. https://doi.org/10.1016/j.envpol.2019.05.084
34. Vuppaladadiyam AK, Yao JG, Florin N, George A, Wang X, Labeeuw L, Jiang Y, Davis RW, Abbas A, Ralph P, Fennell PS, Zhao M (2018) Impact of flue gas compounds on microalgae and mechanisms for carbon assimilation and utilization. ChemSusChem 11(2):334–355. https://doi.org/10.1002/cssc.201701611
35. Yen HW, Ho SH, Chen CY, Chang JS (2015) CO_2, NOx and SOx removal from flue gas via microalgae cultivation: a critical review. Biotechnol J 10(6):829–839. https://doi.org/10.1002/biot.201400707
36. Kong W, Shen B, Lyu H, Kong J, Ma J, Wang Z, Feng S (2021a) Review on carbon dioxide fixation coupled with nutrients removal from wastewater by microalgae. J Clean Prod 292. https://doi.org/10.1016/j.jclepro.2021.125975
37. Van Den Hende S, Vervaeren H, Boon N (2012) Flue gas compounds and microalgae: (Bio-) chemical interactions leading to biotechnological opportunities. Biotechnol Adv 30(6):1405–1424. https://doi.org/10.1016/j.biotechadv.2012.02.015
38. Qie F, Zhu J, Rong J, Zong B (2019) Biological removal of nitrogen oxides by microalgae, a promising strategy from nitrogen oxides to protein production. Bioresource Technology 292. https://doi.org/10.1016/j.biortech.2019.122037
39. Du K, Wen X, Wang Z, Liang F, Luo L, Peng X, Xu Y, Geng Y, Li Y (2019) Integrated lipid production, CO_2 fixation, and removal of SO_2 and NO from simulated flue gas by oleaginous Chlorella pyrenoidosa. Environ Sci Pollut Res 26(16):16195–16209. https://doi.org/10.1007/s11356-019-04983-9
40. Bombelli P, Bradley RW, Scott AM, Philips AJ, McCormick AJ, Cruz SM, Anderson A, Yunus K, Bendall DS, Cameron PJ, Davies JM, Smith AG, Howe CJ, Fisher AC (2011) Quantitative analysis of the factors limiting solar power transduction by Synechocystis sp. PCC 6803 in biological photovoltaic devices. Energy Environ Sci 4(11):4690–4698. https://doi.org/10.1039/c1ee02531g
41. Pisciotta JM, Zou Y, Baskakov IV (2010) Light-dependent electrogenic activity of cyanobacteria. PLoS ONE, 5(5). https://doi.org/10.1371/journal.pone.0010821
42. Fischer WW, Hemp J, Johnson JE (2016). Evolution of oxygenic photosynthesis. Annu Rev Earth Planet Sci 44:647–683. https://doi.org/10.1146/annurev-earth-060313-054810
43. Lakshmidevi R, Gandhi NN, Muthukumar K (2020) Carbon neutral electricity production from municipal solid waste landfill leachate using algal-assisted microbial fuel cell. Appl Biochem Biotechnol 191(2):852–866. https://doi.org/10.1007/s12010-019-03160-5
44. Yang Z, Nie C, Hou Q, Zhang L, Zhang S, Yu Z, Pei H (2019) Coupling a photosynthetic microbial fuel cell (PMFC) with photobioreactors (PBRs) for pollutant removal and bioenergy recovery from anaerobically digested effluent. Chem Eng J 359:402–408. https://doi.org/10.1016/j.cej.2018.11.136

45. Raja SW, Thanuja KG, Karthikeyan S, Marimuthu S (2022) Exploring the concurrent use of microalgae Coelastrella sp. for electricity generation and dairy wastewater treatment. Bioresource Technology Reports 17. https://doi.org/10.1016/j.biteb.2021.100889
46. Maity JP, Hou CP, Majumder D, Bundschuh J, Kulp TR, Chen CY, Chuang LT, Nathan Chen CN, Jean JS, Yang TC, Chen CC (2014) The production of biofuel and bioelectricity associated with wastewater treatment by green algae. Energy 78:94–103. https://doi.org/10.1016/j.energy.2014.06.023
47. Zhang Y, Noori JS, Angelidaki I (2011) Simultaneous organic carbon, nutrients removal and energy production in a photomicrobial fuel cell (PFC). Energy Environ Sci 4(10):4340–4346. https://doi.org/10.1039/c1ee02089g
48. Behl K, SeshaCharan P, Joshi M, Sharma M, Mathur A, Kareya MS, Jutur PP, Bhatnagar A, Nigam S (2020) Multifaceted applications of isolated microalgae Chlamydomonas sp. TRC-1 in wastewater remediation, lipid production and bioelectricity generation. Bioresource Tech 304. https://doi.org/10.1016/j.biortech.2020.122993
49. An J, Li N, Wan L, Zhou L, Du Q, Li T, Wang X (2017) Electric field induced salt precipitation into activated carbon air-cathode causes power decay in microbial fuel cells. Water Res 123:369–377. https://doi.org/10.1016/j.watres.2017.06.087
50. Oh SE, Kim JR, Joo JH, Logan BE (2009) Effects of applied voltages and dissolved oxygen on sustained power generation by microbial fuel cells. Water Sci Technol 60(5):1311–1317. https://doi.org/10.2166/wst.2009.444
51. Sun M, Zhai LF, Li WW, Yu HQ (2016) Harvest and utilization of chemical energy in wastes by microbial fuel cells. Chem Soc Rev 45(10). https://doi.org/10.1039/c5cs00903k
52. Luo H, Yu S, Liu G, Zhang R, Teng W (2016) Effect of in-situ immobilized anode on performance of the microbial fuel cell with high concentration of sodium acetate. Fuel 182:732–739. https://doi.org/10.1016/j.fuel.2016.06.032
53. González Del Campo A, Cañizares P, Rodrigo MA, Fernández FJ, Lobato J (2013) Microbial fuel cell with an algae-assisted cathode: a preliminary assessment. J Power Sources 242:638–645. https://doi.org/10.1016/j.jpowsour.2013.05.110
54. Chen QK, Xiang XH, Yan, P, Liu SY (2024) Enhancing strategies of photosynthetic hydrogen production from microalgae: differences in hydrogen production between prokaryotic and eukaryotic algae. Bioresour Technol 406. https://doi.org/10.1016/j.biortech.2024.131029
55. Kong W, Shen B, Lyu H, Kong J, Ma J, Wang Z, Feng S (2021b) Review on carbon dioxide fixation coupled with nutrients removal from wastewater by microalgae. J Clean Prod 292. https://doi.org/10.1016/j.jclepro.2021.125975
56. Zhou W, Ruan R (2014) Biological mitigation of carbon dioxide via microalgae: recent development and future direction. Scientia Sinica Chimica 44(1). https://doi.org/10.1360/032013-256
57. Neethu B, Das S, Ghangrekar MM (2023) Photosynthetic microbial fuel cell, Biophotovoltaic cell, and microbial carbon-capture cell. Microb Electrochem Technol: Fundam Appl 2:495–527
58. Nur MMA, Dewi RN (2024) Opportunities and challenges of microalgae in biocement production and self-repair mechanisms. Biocatal Agric Biotechnol 103048
59. Xu P, Li J, Qian J, Wang B, Liu J, Xu R, Chen P, Zhou W (2023) Recent advances in CO_2 fixation by microalgae and its potential contribution to carbon neutrality. Chemosphere 319. https://doi.org/10.1016/j.chemosphere.2023.137987
60. Clarens AF, Nassau H, Resurreccion EP, White MA, Colosi LM (2011) Environmental impacts of algae-derived biodiesel and bioelectricity for transportation. Environ Sci Technol 45(17):7554–7560. https://doi.org/10.1021/es200760n
61. Lakshmidevi R, Nagendra Gandhi N, Muthukumar K (2022) Bioelectricity and bioactive compound productionin an algal-assisted microbial fuel cell with immobilized bioanode. Biomass Conversion and Biorefinery 12(8):3457–3473
62. Zhou M, He H, Jin T, Wang H (2012) Power generation enhancement in novel microbial carbon capture cells with immobilized Chlorella vulgaris. J Power Sources 214:216–219
63. Lam MK, Lee KT (2012) Immobilization as a feasible method to simplify the separation of microalgae from water for biodiesel production. Chem Eng J 191:263–268. https://doi.org/10.1016/j.cej.2012.03.013

64. de-Bashan LE, Bashan Y (2010) Immobilized microalgae for removing pollutants: review of practical aspects. Bioresource Tech 101(6):1611–1627. https://doi.org/10.1016/j.biortech.2009.09.043
65. Han M, Zhang C, Ho S-H (2022) Immobilized microalgal system: an achievable idea for upgrading current microalgal wastewater treatment. Environ Sci Ecotechnology 14:100227. ISSN 2666-4984. https://doi.org/10.1016/j.ese.2022.100227
66. Ng F-L, Phang S-M, Periasamy V, Yunus K, Fisher AC (2017) Enhancement of power output by using alginate immobilized algae in biophotovoltaic devices. Sci Rep 7(1):16237. https://doi.org/10.1038/s41598-017-16530-y
67. Hadiyanto H, Christwardana M, Zuli PW, Purwanto P, Sudarno S, Haryani K, Hoang AT (2022) Response surface optimization of microalgae microbial fuel cell (MMFC) enhanced by yeast immobilization for bioelectricity production. Chemosphere 287: 132275, Part 3. ISSN 0045-6535. https://doi.org/10.1016/j.chemosphere.2021.132275
68. Yagishita T, Sawayama S, Tsukahara K, Ogi T (2000) Behavior of bioelectrocatalytic currents in electrolysis cells using immobilized Synechocystsis sp. PCC6714 within poly-ion complex on glassy carbon electrodes. Electrochemistry 68(11):878–881
69. Yagishita T, Sawayama S, Tsukahara KI, Ogi T (1998) Performance of photosynthetic electrochemical cells using immobilized Anabaena variabilis M-3 in discharge/culture cycles. J Ferment Bioeng 85(5):546–549
70. Alami AH, Alasad S, Ali M, Alshamsi M (2021) Investigating algae for CO_2 capture and accumulation and simultaneous production of biomass for biodiesel production. Sci Total Environ 759. https://doi.org/10.1016/j.scitotenv.2020.143529
71. Ng FL, Phang SM, Periasamy V, Yunus K, Fisher AC (2014a) Evaluation of algal biofilms on Indium Tin Oxide (ITO) for use in biophotovoltaic platforms based on photosynthetic performance. PLoS ONE 9(5). https://doi.org/10.1371/journal.pone.0097643
72. Periasamy CH, Pindah V, Yunus JK, Fisher AC (2021) Integration of bioelectricity generation from algal biophotovoltaic (BPV) devices with remediation of Palm Oil Mill Effluent (POME) as substrate for algal growth. Environ Technol Innov 21. https://doi.org/10.1016/j.eti.2020.101280
73. Qi X, Liu X, Gu Y, Liang P (2024c) Whole-cell biophotovoltaic systems for renewable energy generation: a systematic analysis of existing knowledge. Bioelectrochemistry 158. https://doi.org/10.1016/j.bioelechem.2024.108695
74. Tay ZHY, Ng FL, Thong CH, Lee CW, Gnana kumar G, Al-Sehemi AG, Phang SM (2024) Evaluation of selected tropical marine microalgal cultures for use in biophotovoltaic platforms. Appl Microbiol Biotechnol 108(1):71
75. Zhou W, Wang J, Chen P, Ji C, Kang Q, Lu B, Li K, Liu J, Ruan, R (2017) Bio-mitigation of carbon dioxide using microalgal systems: advances and perspectives. In renewable and sustainable energy reviews, vol 76, pp 1163–1175. Elsevier Ltd. https://doi.org/10.1016/j.rser.2017.03.065
76. He H, Zhou M, Yang J, Hu Y, Zhao Y (2014) Simultaneous wastewater treatment, electricity generation and biomass production by an immobilized photosynthetic algal microbial fuel cell. Bioprocess Biosyst Eng 37(5):873–880.
77. Ng FL, Phang SM, Thong CH, Periasamy V, Pindah J, Yunus K, Fisher AC (2021) Integration of bioelectricity generation from algal biophotovoltaic (BPV) devices with remediation of palm oil mill effluent (POME) as substrate for algal growth. Environ Technol Innov 21:101280

Algae-Based Proteins and Biomass as Clean-Label Ingredients for the Food Industry

Juliana Botelho Moreira, Lisiane Fernandes de Carvalho, José Matheus, Catarina Prista, and Anabela Raymundo

Abstract The growing consumer concern about health has increased interest in food products free of additives, driving the clean-label concept. Algae biomass can offer several compounds for this context, including proteins and pigments. Algal proteins are recognized for their high biological value, demonstrating several health benefits. Proteins and other components constituted in algae biomass contribute to achieving desired structural and rheological properties in food matrices. Microalgae pigments such as phycobiliproteins, carotenoids, and chlorophylls can replace artificial colorings used in the food industry. Besides, microalgae proteins have the potential for the development of innovative materials to be applied in food packaging. Thus, using algae as a protein source in food products becomes an approach with different functionalities for the clean-label trend. The possibility of promoting nutrition and health while coloring and preserving food make algae protein an ingredient with high innovation potential for the food industry. This review article focuses on microalgae and macroalgae as a source of proteins and biomass for use as a natural ingredient within the clean-label trend. This review also brings the potential of proteins in packaging and coatings and the main challenges in using algae biomass as a clean-label ingredient.

Keywords Microalgae biomass · Nutrition · Seaweed extracts · Sustainability

1 Introduction

Consumers currently favor natural and sustainably produced foods over ultra-processed processed alternatives, motivated by apprehensions about artificial additives. As the clean-label trend progresses, consumer standards are shifting. There

J. B. Moreira (✉) · L. F. de Carvalho · J. Matheus · C. Prista · A. Raymundo
LEAF—Linking Landscape, Environment, Agriculture and Food Research Center, Associated Laboratory TERRA, Instituto Superior de Agronomia, Universidade de Lisboa, Tapada da Ajuda, Lisboa, Portugal
e-mail: julianapvi@yahoo.com

A. Khan and K. A. Alzahrani (eds.), *Microalgae Biotechnology and Sustainable Development*, Sustainable Materials and Technology,
https://doi.org/10.1007/978-981-95-2597-3_3

is a rising trend toward adopting a "clean-label" processing method to improve product preservation and tackle sustainability issues [52]. Clean labeling extends beyond mere label designations and pertains to foods' healthiness and sustainability [90]. The clean-label concept primarily revolves around the incorporation of "natural" ingredients, which are readily identifiable and deemed safe by consumers [56]. Although not mandated by food regulations, clean-label food items boast streamlined ingredient lists devoid of artificial additives, aligning with the increasing focus on health awareness and environmental responsibility within the food sector.

Algae derivatives find extensive commercial utilization, ranging from food supplements in powders and oils to ingredients and additives [128]. The integration of algae biomass facilitates the development of innovative food and ingredient products [89, 93, 129], serving as a significant reservoir of bioaccessible proteins [21] and accentuating their value in various applications. Microalgae boast a remarkable protein content (~60%), surpassing other protein-rich foods like meat, yeast, eggs, and soy. The cyanobacteria *Spirulina* can accumulate up to 70% of proteins in its biomass [81]. Additionally, microalgae exhibit diverse colors attributed to pigments produced in their chloroplasts [96]. These pigments, notably chlorophylls, carotenoids, and phycobiliproteins, not only serve as natural food dyes but also offer health benefits owing to their bioactivity [81]. Therefore, besides providing natural color to food products in the clean-label context, using microalgae as a colorant also sustainably promotes health.

Moreover, incorporating seaweed extracts, microalgae biomass, or even their polysaccharides, such as agar, carrageenan, and alginate, into food products can enhance their rheological properties and functionality. [9], for instance, explored how various seaweeds affect the properties of raw earth material and found that adding freeze-dried *Solieria chordalis* and *Sargassum muticum*, or their polysaccharides, significantly boosted compressive strength by threefold and improved water resistance. Other work discovered that aqueous suspensions of fine particles of the seaweed *Durvillaea antarctica* showed similar behavior to two commercial thickeners, being suitable for various applications [30]. Furthermore, the addition of *Spirulina platensis* improves the rigidity and compactness of soy protein isolates without altering its tertiary structure, demonstrating that this cyanobacterium enhances the rheology and mechanical properties of food while also serving as a nutritional supplement [131]. These studies show that using algae in foods can potentially reduce the necessity for numerous food additives, aligning with the clean-label concept.

Compounds from algae also show potential in developing antimicrobial packaging materials that extend food shelf life [80] and intelligent packaging [122], improving food safety and food security. Besides, in carrageenan-based edible films produced from seaweed extracts, the mechanical properties and viscosity are enhanced, suggesting the potential of seaweed-based additives to improve the functionality of food packaging films [3]. Thus, algae, known for their high protein and polysaccharide content, are promising approaches for sustainable packaging innovations and overcoming challenges linked to industrial processes and the rheological properties of food packaging materials. In this way, this review article focuses on microalgae and macroalgae as a source of proteins and biomass for use as a natural ingredient within the clean-label trend. This review also brings the potential of proteins in packaging and coatings and the main challenges in using algae biomass as a clean-label ingredient.

2 Food Industry and Clean-Label Concept

Consumer food preferences have focused on nutritional aspects, health benefits, and food safety. These preferences are related to eating habits influenced by cultural, political, socioeconomic, and demographic factors. In addition to the search for more natural foods, extending the shelf life of products and low environmental impact are consumer demands that have been growing constantly [126]. Thus, since consumers have shown interest in sustainable products that do not contain synthetic additives, the clean-label concept has become increasingly popular [22, 51]. Therefore, the clean-label trend is an outstanding strategy for food industries to attract consumers (Fig. 1).

The Institute of Food Technologists defines a clean-label product as one crafted with a minimal number of ingredients, including those that customers can comprehend and perceive as healthy [29]. This term typically refers to products with a short list of ingredients, as well as free of synthetic additives, such as artificial flavors, colors, and preservatives that hurt health [22, 126]. Given the diversity of ingredients and food categories, understanding consumers' perception of individual ingredients

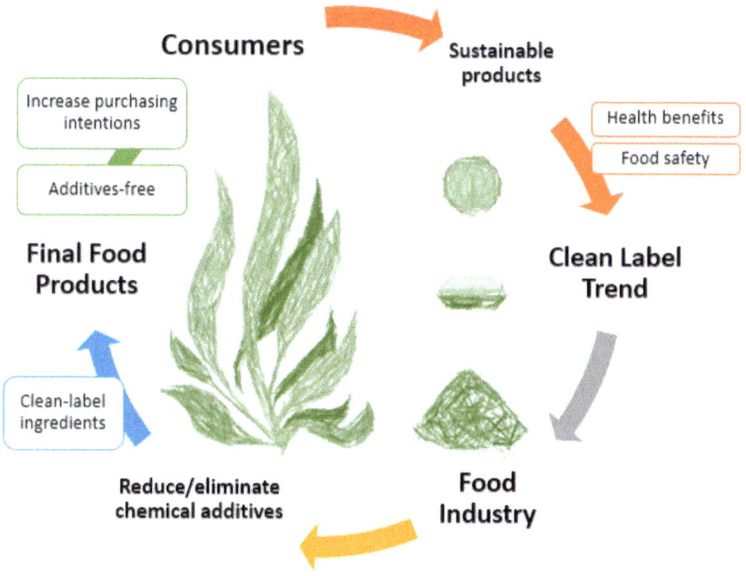

Fig. 1 Illustrative representation of the clean-label trend for consumers

on labels is a crucial factor for driving the entry of new ingredients and the launch of new products free of synthetic additives into the food market [7].

Increasing consumer's knowledge of food ingredients can improve their perceived naturalness and acceptability in the clean-label movement [69]. Given that clean-label foods lack a legal definition, interpretations can vary. However, it is commonly accepted that such products contain natural ingredients and avoid over-processed compounds [4]. In this sense, [38] observed that Portuguese people were more willing to eat and buy the clean-label versions of all products offered.

Thus, food industries seek to evolve in research and adapt their technologies for fabricating new products added with clean-label ingredients to meet consumer needs while contributing to sustainability. Natural additives that claim to be natural antimicrobial preservatives and antioxidants are an attractive field of research. In addition, the perception of protein as a clean-label ingredient is of particular interest due to the current relevance in developing alternative protein foods [7].

3 Algae for Foods

Algae are photosynthetic and aquatic organisms, where the different species can be from single-celled microalgae or in a multicellular form known as macroalgae (seaweed). Algae have great potential for mitigating carbon dioxide emissions, and their cultivation is economical and efficient as they can produce food throughout the year and do not compete with food crops because they do not require arable land. They effectively contribute directly or indirectly to achieving the 17 Sustainable Development Goals (SDGs), which are part of an action plan for a better and more sustainable world by 2030 [43, 95].

Algae have been part of human nutrition for centuries and possess a high protein content with excellent amino-acid profile. Algae are also a good source of polyunsaturated fatty acids, pigments, vitamins, and minerals [15, 55]. The use of algae in foods brings beneficial effects to health, including antioxidant, anti-inflammatory, antimicrobial, antiviral, anticoagulant, anticancer, and antihypertensive activities [26]. Therefore, healthier and more sustainable foods can be developed from microalgae and macroalgae.

To supply people with necessary nutrients, microalgae and macroalgae can be consumed directly as food or indirectly through their use as feed in the animal agriculture industry [88, 106] and agriculture as soil fertilizers [47]. Consequently, to meet the needs of consumers for nutritious, sustainable, affordable, tasty, and healthy food, the incorporation of algae in a variety of food products has been extensively studied, and their consumption has increased throughout the world [33, 50, 66, 71].

The global algae market presents a diversity of algae products. Macroalgae products, mainly in the form of hydrocolloids, have an annual market of approximately 6.7 billion dollars. Seaweed used in Asian cuisine, such as nori produced from dried *Porphyra* leaves, presents a market of 1 billion dollars per year. Microalgae, about dry biomass, have values exceeding 1.25 billion dollars per year [125].

There are between 200,000 and 800,000 species of microalgae in nature. *Chlorella* genera and *Arthrospira platensis* (better known as Spirulina) are the most exploited in the food sector. Due to their high protein content, they are cultivated as an alternative source of protein. Genus such as *Dunaliella, Isochrysis, Nannochloropsis,* and *Haematococcus* have been investigated for incorporation into foods [42, 62, 76, 114, 134, 137]. Due to their bioactive compounds, microalgae and macroalgae can contribute to developing functional and innovative food products. Furthermore, the nutritional potential and safety of different species must be investigated to stimulate the commercialization of algae biomass [22, 88, 94].

It is known that the genus *Arthrospira* has a long worldwide history of use as a food source, supplement, and food additive. For commercial and marketing applications, instead of using *Arthrospira,* Spirulina is used, a name that makes it easier to pronounce and helps to make it easier to remember. The rich chemical composition and the broad potential for application of Spirulina in food sectors have instigated researchers to seek a more detailed analysis of the species of this genus [92]. Nowicka et al. [92] concluded that the type of species of the genus *Arthrospira* based on a polyphasic approach revealed differences between *A. jenneri* and organisms produced under the common name *Spirulina.* With this, the authors proposed a new genus called *Limnospira,* encompassing species with high industrial utilization and potential.

4 Algae as Future in Clean-Label Products

The growing trend of "clean-label" foods focuses on ensuring these foods are safe for consumption by replacing synthetic additives [49], which could potentially lead to an increase in the consumption of algae biomass. Incorporating algae into food can induce beneficial alterations in the food products' physical, chemical, biological, and structural properties [74, 97]. In addition to the array of biochemical characteristics algae bring, their incorporation into the food industry significantly impacts rheological aspects. This fact is particularly relevant because the algae composition of macronutrients such as proteins, fats, and polysaccharides directly influences the microstructure of food products, including the gel network, eliminating the need for multiple thickeners and stabilizers [99, 100]. Seaweeds also significantly influence the alteration of rheology properties, principally if undergoing thermal processing through hydration that releases the gelling agents [12], known as hydrocolloids. In this scenario, by incorporating algae into food formulations, it becomes possible to create cleaner labels, thanks to its capability to replace other ingredients [36] through the diverse biomass traits it offers, such as in the case of incorporating *Chlorella vulgaris* in frankfurters, which exhibited antibacterial effects, and helped to extend product shelf life during chilled storage [20].

Seaweeds, for instance, despite being consumed for centuries in Eastern history, are on the rise worldwide with great applicability to the clean-label trend. In industry, they are mainly known for their hydrocolloids (carrageenan, alginate, and agar), which consist of polysaccharides characterized by a complex structure serving as thickening, emulsifying, gelling, and/or stabilizing agents [140, 17]. By incorporating whole algae biomass [135], the nutritional benefits provided by proteins and other compounds are taken advantage of while providing advantages with rheological aspects of this raw material, ensuring a cleaner product. Besides, macroalgae commonly serve as a notable reservoir of glutamate, facilitating the "umamification" of the matrices in which they are incorporated [86] while encouraging reduced salt intake.

Microalgae, on the other hand, if compared to conventional animal- or plant-based sources, own some additional advantages like higher growth rate, less water demand, and adaptation to different environmental conditions [15], contributing to sustainable climate practices and following SDGs [127]. They are more than natural and environmentally friendly food colorants [88], and their biomass may be used for several applications.

Achieving a precise balance between sustainability practices and the clean-label movement while maintaining the desired taste, texture, aroma, and appearance of food products is not just a goal but a necessity in today's food industry. By embracing innovation, transparency, and consumer education, food producers meet these demands and pave the way for a more ethical, environmentally friendly, and enjoyable culinary landscape for generations to come [13, 34, 104].

5 Application of Micro/Macroalgae in Food Development

The use of algae biomass (seaweed, microalgae, or even cyanobacteria, which are constantly informally disseminated as microalgae), such as *Gelidium corneum*, *Chlorella vulgaris*, and *Spirulina* sp. (Fig. 2), and many others, is increasingly common in food products, such as cereal or legume-based products, including couscous [57], pasta [39, 107], hummus [70], cookies [11], fermented dairy products [50], and snacks made by 3D printing [97].

Other food products added with Spirulina have been developed, such as beverages [89], snacks [62, 65], food supplements for athletes [24], shakes for the elderly population [111], energy gels [79], and yogurts [136]. Seaweeds, such as kombu and nori,

Fig. 2 Algae powder. From left to right: Biomass from seaweed *Gelidium corneum*, microalgae *Chlorella vulgaris*, and cyanobacteria *Arthrospira sp*

surpassed the barriers of Japanese cuisine. *Himanthalia elongata* has been incorporated into frankfurters [64], *Lamina japonica* into sausages [58], *Fucus vesiculosus* to fish products [132], *Sargassum marginatum* and *Undaria Pinnatifida* to pasta, and *Laminaria* into yogurt, cheese, and milk deserts [112].

The incorporation of algae into food products is meeting with less skepticism than it was years ago. The 2024 Innova Market Insights survey identifies "Oceans of Possibilities" as a significant trend, emphasizing algae in product formulations. This trend supports marine conservation and the development of new plant-based ingredients, positioning oceans as future farms for sustainable resources. Food and beverage launches with microalgae have grown over 40% annually in recent years, with Spirulina becoming a popular environmentally friendly ingredient [53].

Due to the rise in popularization that its benefits are obtaining, such as for health and sustainability, the time has come to give visibility and applicability to these elements from the sea [85], targeting food products with innovative colors, health, and nutrition. Previous reports indicate the reasons for the limited incorporation of microalgae into commercial products. The low concentration, generally under 1%, is used due to the intense green color of microalgae being a significant drawback [61]. Furthermore, another high barrier to incorporating algae into foods is the aroma, generally marked by fishy or sea flavor. If these foods are not flavored, consumers will not be willing to pay for them since flavor is the main element to consuming products of this nature, followed by the nutrient content and the health benefits [138].

The promotion of veganism by eliminating animal-based foods and adopting sustainable and health-conscious dietary choices has provided a significant avenue for exploring alternative protein sources in human nutrition [84]. Algae holds vast potential as a source of sustainable and high-quality proteins for human consumption [15]. Regarding resilient food, seaweeds are promising candidates to avoid a global famine [54], and microalgae can be used as nutritional added value commodities in the near future [115]. Moreover, most marine food products and by-products are being converted and utilized as valuable functional foods through recent developments in the marine bioprocess industry [117].

Therefore, the introduction of algae into food products presents an exciting prospect. The nutritional richness and flavor-enhancing properties of algae have a high potential to add depth and diversity to culinary experiences [6], depending on the experience. Therefore, it is essential to evaluating the combinations of microalgae with other ingredients to obtain surprising final products that easily engage the consumer. As ongoing research explores their applications, algae become essential approaches in shaping the landscape of food innovation and consumer preferences.

5.1 Micro and Macroalgae as an Emerging Protein Source

The global population continues to grow and is expected to reach 9.7 billion people by 2050. Furthermore, environmental issues are becoming increasingly important. In addition to a sharp increase in population growth, the scarcity of arable land, water, biodiversity, and mainly protein sources turns the goal of feeding the world population in the near future into an emerging challenge. It is time to rethink our diet, and algae should play a relevant role in this framework [15, 26, 40].

Finding more sustainable food sources to meet the increase in protein demand is crucial. Algae have advantages in protein production, such as sustainable cultivation, protein yield, and nutritional quality, which makes these alternative protein sources in food development [10, 71]. Also, the quality and content of algae protein, for example, *Arthrospira platensis*, which can reach close to 70%, stand out concerning traditional protein sources such as soy (37%), milk (26%), yeast (39%), dried eggs (47%), and meat (43%) [84].

In addition, algae consumption has been increasing due to the popularity of vegan and vegetarian diets and plant-based foods worldwide [26]. *Chlorella* and *Arthrospira* are the most sold on the market. *Arthrospira platensis* is eaten worldwide as a nutritional supplement in powder, flakes, capsules, or as a colorant. The protein ranges are around 17–73%, and for the amino-acid profile, it has a high content of isoleucine, valine, lysine, threonine, phenylalanine, and tyrosine. *Chlorella vulgaris* is a microalga with a content of protein which can vary in the range 7–70% and with a higher essential amino-acid content in leucine, valine, lysine, threonine, phenylalanine, and tyrosine [31, 40, 60, 87]. These protein contents in microalgae and macroalgae vary due to several factors such as species used and cultivation conditions, climate, and production location. Given this, these factors can be altered to stimulate the increase of proteins. Therefore, it is necessary to carry out more research to boost their use in human food applications [59, 94, 125].

The level of microalgae incorporation in food depends on the technological impact on the food structure but mainly on the final product's sensory profile. Generally, it is not feasible to incorporate high levels of biomass. Table 1 shows the levels of microalgae incorporation and the protein content of some products added with microalgae and macroalgae species.

Seaweed contains levels of protein ranging from 5 to 40% dry weight and with a large fraction of amino acids such as glycine, alanine, taurine, arginine, proline, glutamic acid, and aspartic acid, which cover the intake requirements established by FAO [48, 133]. However, it is crucial to evaluating the protein digestibility to estimate its nutritional quality. Digestibility varies among seaweed species and is also subject to seasonal variations in the chemical composition, including changes in glycoprotein and antinutritional factors [124].

5.2 Potential of Phycobiliproteins as a Clean-Label Ingredient

Food color is a quality parameter that influences consumers' acceptance of products. There has been a growing concern about using artificial colorings due to health reasons. As a result, there is a rising demand for natural colorings in the food industry

Table 1 Protein content of the products with the addition of various microalgae and macroalgae species

Species	Biomass (%)	Protein content (%)	Products with increased protein content	References
Chlorella vulgaris	2–4 (w/v) 2–4 (w/v) 1.9 (w/w)	10.1–11.2 5.4–5.9 11.83	Cream cheese Quark cheese Bread	Falco et al. [37] Falco et al. [37] Pereira et al. [100]
	4 (w/w)	7.29	Gluten-free bread	Qazi et al. [103]
Arthrospira platensis	12.5 (w/v) 2–6 (w/w) 0.8 (w/w) 0.5 (w/w)	51.32 6.1–7.8 15.75 4.17	Vegetal soybean drink Cookies Cheese Yogurt	Niccolai et al. [89] Batista et al. [11] Mazinani et al. [73] Priyanka et al. [102]
Dunaliella salina	1–3 (w/w)	13.35–13.60	Pasta	El-baZ et al. [35]
Tetraselmis chuii	4 (w/w)	7.16	Gluten-free bread	Qazi et al. [103]
Nannochloropsis gaditana	4 (w/w)	7.29	Gluten-free bread	Qazi et al. [103]
Isochrysis galbana	2 (w/w)	4.0	Yogurt	Matos et al. [72]
Undaria pinnatifida	5–30 (w/w)	18.92–21.68	Pasta	Prabasankar et al. [101]
Ascophyllum nodosum	0.05 (w/w)	14.42	Pork liver pâtés	Agregan et al. [2]
Fucus vesiculosus	0.05 (w/w)	14.17	Pork liver pâtés	Agregan et al. [2]
Kappaphycus alvarezzi	15–25 (w/w)	8.5–13.2	Instant spice	Senthil et al. [113]
Enteromorpha compressa	5–15 (w/w)	15.6–20.0	Pakoda	Mamatha et al. [67]

[105]. Algae are a valuable source of natural pigments such as carotenoids, chlorophylls, and phycobiliproteins recognized for their coloring potential in addition to their beneficial properties for health [68].

Phycobiliproteins are fluorescent hydrophilic proteins characterized by having several chromophores called phycobilin attached to their polypeptide structure [32]. Besides, they are water-soluble proteins and constitute up to 60% of the soluble protein in algae cells. Phycobiliproteins' classification includes phycoerythrin, allophycocyanin, and phycocyanin [8]. Phycocyanin is the major phycobiliprotein, followed by phycoerythrin and allophycocyanin [120]. Phycocyanin is the first organic blue protein pigment approved in the USA, Europe, and Asia [60]. Phycocyanin is mainly extracted from *Arthrospira* species, which can contain up to 14% (w/w) of this pigment in total cell protein at an industrial production scale [79]. Ice creams [5, 19, 108], yogurts [75, 110], and candies and gummies [8, 63] are some food products where phycocyanin has been incorporated. In addition to contributing to the color aspects of foods, phycocyanin has antioxidant properties for use as an anti-browning agent in food preservation [139]. Phycocyanin's antimicrobial activity against gram-negative and gram-positive bacteria is also reported [116], intensifying its capacity for food preservation. Therefore, phycocyanin has the potential to be used as a clean-label ingredient, not only for replacing synthetic dyes but also for preserving food products, eliminating the use of artificial chemical preservatives.

Phycoerythrin is found in most red microalgae Cyanophyceae and Cryptophyceae, and macroalgae. For example, the red color of Phylum Rhodophyta is associated with phycoerythrin [45]. Some studies have evaluated phycoerythrin in food applications, such as ice creams [41] and milk-based products [44]. Although phycoerythrin has been most investigated for pharmaceutical applications due to antioxidant and anti-inflammatory properties, this natural pigment could expand the list of natural coloring additives available on the food market [44]. Allophycocyanin absorbs red light and transfers the energy to the reaction centers of oxygenic photosynthesis [46]. These phycobiliproteins are found in blue-green and red algae and present potential for use as a food colorant [68]. However, the direct application of this pigment in foods has not yet been demonstrated in research studies.

Therefore, the phycobiliprotein that stands out in this scenario in terms of the potentiality for use as a clean-label ingredient is still phycocyanin. The other phycobiliproteins also have a high potential to be utilized as natural food colorants to enhance the sensory attributes of the end product while also providing nutritional value and preservation of foods. Moreover, incorporating phycobiliproteins in food products drives a new segment for algae-based compounds in the clean-label concept for the food industry.

Microalgae-based biorefinery is an innovative and sustainable approach that leverages algae-based biomass to recover multiple products through integrated processes [130]. The biorefinery allows the utilization of residues generated from extraction processes (the vast majority of biomass) in various applications. The residual biomass of phycocyanin extraction, rich in proteins and fatty acids, can be used to develop food products in a biorefinery concept. In this context, [18] developed a vegan emulsion containing residual biomass from Spirulina phycocyanin extraction. The study concluded that it is possible to apply a biorefinery approach allowing the incorporation of an innovative clean-label ingredient. This ingredient can be used in food products as a protein source and enhances health benefits due to its antioxidant activity. According to the authors, this clean-label strategy would appeal to consumers who seek vibrant-colored products aligned with sustainability and circular economy concepts.

In addition to its potential in food products, microalgae biorefinery was applied to extract and purify phycocyanin from *Arthrospira platensis* and produce biocrude [109]. Another work focused on microalgae sulfate remediation, where the generated biomass was used for sequential phycocyanin extraction and the crude lipid extract for further valorization into biofuels [16]. Considering the importance of increasing the economic viability of phycocyanin production, this residual biomass from the microalgae biodiesel production can still be used as a biofertilizer [27] Besides, residual biomass of several extraction processes can feed anaerobic digesters for biogas production [27]. Thus, microalgae-based biorefinery is a promising strategy to increase the economic viability of large-scale production of proteins and phycocyanin. In addition to contributing to environmental sustainability, the microalgae biorefinery approach maximizes resources and promotes a circular bioeconomy in which waste can become a valuable input for future production cycles of high-value products [28, 130].

6 Perspectives of Microalgae Proteins Applied in Food Packaging and Coatings

Food packaging materials and coating made from biopolymers should meet several criteria to ensure health safety, mechanical strength, thermal resistance, and barrier properties. Additionally, their durability and functionality are also important, apart from their ability to biodegrade [1]. Proteins present the potential to be inserted in this context, in addition to standing out due to their relative abundance, good film-forming ability, and high nutritional value. Besides, protein-based polymers are the most useful due to their excellent gas barrier and mechanical properties compared to polysaccharides and lipids [14]. Therefore, the high protein content of macroalgae and microalgae suggests their use for manufacturing packaging materials and edible coatings.

Edible packaging has been developed using proteins and polysaccharides that are safe for human consumption, resulting in zero waste [23]. The use and marketing of algae compounds for food or edible packaging purposes has gained considerable attention in recent years. This field is regulated by several organizations, including the Food and Drug Administration (FDA), the European Parliament and Council, the Joint World Health Organization, and the Food and Agricultural Organization Expert Committee on Food Additives [119]. Algae are a significant source of bioactive molecules. When added to protein-based films and coatings, phenolic compounds can improve the mechanical, physical, and functional properties of these materials [23].

Innovative technologies have been investigated to develop protein-based food packaging with interesting consumer responses. The electrospun protein membranes can be transformed into small components such as labels, sachets, and stickers. These can be applied to primary packaging to create an intelligent or active function in food packaging [78, 79]. Furthermore, the use of phycobiliproteins such as phycocyanin from Spirulina is being expanded into intelligent food packaging to observe the freshness of food through its application as a pH indicator [83, 121, 122] and into active packaging to preserve foods [80].

7 Challenges for Using Algae in Clean-Label Food Products

The clean-label trend contributes to diverse opportunities for the algae production industries, offering possibilities to improve foods' nutritional, technological, and sensory properties. However, achieving the same shelf life without using chemical preservatives and keeping products safe and attractive to consumers is also a challenge for developing clean-label products. The main obstacle in using natural antioxidants and antimicrobials from macroalgae and microalgae is achieving efficacy comparable to synthetic additives. The most significant challenge lies in determining the appropriate level of incorporation to the compound which is effective and does not compromise the food's quality characteristics, such as flavor, aroma, and color [52]. Phycocyanin, for example, is recognized for its antioxidant activity and intense blue color. However, it may present some application limitations in food preservation in foods where the blue tone is unsuitable.

In general, the intense flavors of algae can limit its utilization as a clean-label ingredient in certain foods. Moreover, bioactive compound loss during processing is another challenge to produce foods containing algae [25]. In this sense, innovative encapsulation techniques have overcome these challenges [82]. Regarding the intense color aspects of microalgae and macroalgae, food products developed in lighter, more aesthetically neutral shades of green are more acceptable to consumers. One strategy for changing the tone of algae biomass is to change the carbon/nitrogen ratio of the cultivation medium. Therefore, it becomes challenging to find the balance between the incorporated biomass content to contribute nutrients and bioactive compounds without compromising the food's physical and sensory properties and quality. A food product can bring several health benefits, however, if it is not tasty and attractive, few consumers will be willing to buy it [81, 91].

Besides, harvesting is one of the most expensive procedures of microalgae production [81]. Around 90% of the cost of obtaining microalgae biomass from open ponds is due to energy-intensive harvesting techniques such as flotation, centrifugation, filtration, and electricity-based techniques [118]. The costs associated with drying are another challenge. The cost of drying *Arthrospira* can amount to 30% of the total production price. Depending on the technique applied, drying can affect the bioactive compounds in microalgae biomass. Conventional drying can cause a loss of phycocyanin of 50–90% in *Arthrospira* biomass [25]. These costs related to processes involving biomass production can be limiting factors in using algae and its compounds as a clean-label ingredient and raw material for application in packaging and films. Therefore, the economic sustainability of microalgae biomass commercial-scale production must be improved so that algae industries can reach their full potential.

According to Celekli et al. [25], the food market that involves using microalgae faces problems with legislation and regulation. The manufacturing processes for algae and its compounds must comply with regulations, and this can increase the overall cost of manufacturing clean-label algae-based ingredients for commercialization. Therefore, legislative measures are essential to ensuring adequate monitoring programs, the correct labeling of food products, and the establishment of safe portion sizes for consumers [81].

8 Conclusions

Integrating algae into food products as clean-label ingredients offers a compelling remedy for various modern-day hurdles in the food sector. Leveraging the nutritional richness of algae proteins enables food producers to bolster product nutrition and satisfy the escalating call for more natural, sustainable, and environmentally friendly options. Adopting algae can diminish reliance on artificial food additives, promoting clearer labeling and healthier dietary preferences. Moreover, the high protein content and bioactive compounds in algae biomass drive the development of new packaging materials and edible coatings. Although there are some challenges regarding production costs, sensory aspects, and legislation, embracing algae and its compounds as a clean-label ingredient represents a significant stride toward forging a more sustainable, eco-conscious food landscape, promising a healthier future for consumers and the environment.

Acknowledgements The authors would like to thank the BLUEWAYSE project—BLUE WAY to a Sustainable Europe, reference SBEP/0004/2023; and the Portuguese Foundation for Science and Technology under the project UIDB/04129/2020 of LEAF Research Center.

References

1. Agarwal A, Shaida B, Rastogi M, Singh NB (2023) Food packaging materials with special reference to biopolymers-properties and applications. Chem. Afr. 6:117–144. https://doi.org/10.1007/s42250-022-00446-w
2. Agregán R, Franco D, Carballo J, Tomasevic I, Barba FJ, Gomez B, Muchenje V, Lorenzo JM (2018) Shelf life study of healthy pork liver pate with added seaweed extracts from Ascophyllum nodosum, Fucus vesiculosus and Bifurcaria bifurcata. Food Res Int 112:400–411. https://doi.org/10.1016/j.foodres.2018.06.063
3. Albertos I, Martin-Diana AB, Burón M, Rico D (2019) Development of functional bio-based seaweed (Himanthalia elongata and Palmaria palmata) edible films for extending the shelf-life of fresh fish burgers. Food Packag Shelf Life 22:100382. https://doi.org/10.1016/j.fpsl.2019.100382
4. Alexandri M, Kachrimanidou V, Papapostolou H, Papadaki A, Kopsahelis N (2022) Sustainable food systems: the case of functional compounds towards the development of clean label food products. Foods 11:2796. https://doi.org/10.3390/foods11182796
5. Amarante MCA, Braga ARC, Sala L, Kalil SJ (2020) Colour stability and antioxidant activity of C-phycocyanin-added ice creams after in vitro digestion. Int Food Res 137:109602. https://doi.org/10.1016/j.foodres.2020.109602
6. Ampofo J, Lord A (2022) Microalgae: bioactive composition, health benefits, safety and prospects as potential high-value ingredients for the functional food industry. Foods 11:1744. https://doi.org/10.3390/foods11121744

7. Aschemann-Witzel J, Varela P, Peschel AO (2019) Consumers' categorization of food ingredients: do consumers perceive them as 'clean label' producers expect? an exploration with projective mapping. Food Qual. Prefer. 71:117–128. https://doi.org/10.1016/j.foodqual.2018.06.003
8. Athiyappan KD, Routray W, Paramasivan B (2024) Phycocyanin from Spirulina: a comprehensive review on cultivation, extraction, purification, and its application in food and allied industries. Food and Humanity 2:100235. https://doi.org/10.1016/j.foohum.2024.100235
9. Autem Y, Bourgougnon N, Kim Guihéneuf S, Perrot A (2023) Comparative study of various seaweed parietal polysaccharides on rheological, mechanical, and water-durability properties of earth-based materials. Mater Struct 56:108. https://doi.org/10.1617/s11527-023-02195-9
10. Barbosa JC, Machado D, Borges S, Pintado M, Gomes AM (2024) Sustainable protein sources: insects, algae and yeast, Editor(s): Geoffrey W. Smithers, Encyclopedia of Food Safety (Second Edition), Academic Press, pp 262–272
11. Batista AP, Niccolai A, Fradinho P, Fragoso S, Bursic I, Rodolfi L, Biondi N, Tredici MR, Sousa I, Raymundo A (2017) Microalgae biomass as an alternative ingredient in cookies: sensory, physical and chemical properties, antioxidant activity and in vitro digestibility. Algal Res 26:161–171. https://doi.org/10.1016/j.algal.2017.07.017
12. Belattmania Z, Bhaby S, Nadri A, Khaya K, Bentiss F, Jama C, Reani A, Vasconcelos V, Sabour B (2021) Gracilaria gracilis (Gracilariales, Rhodophyta) from Dakhla (Southern Moroccan Atlantic Coast) as source of agar: Content, chemical characteristics, and gelling properties. Mar Drugs 19:672. https://doi.org/10.3390/md19120672
13. Bhagat R, Walia SS, Sharma K, Singh R, Singh G, Hossain A (2024) The integrated farming system is an environmentally friendly and cost-effective approach to the sustainability of agri-food systems in the modern era of the changing climate: a comprehensive review. Food Energy Secur. 13:e534. https://doi.org/10.1002/fes3.534
14. Bhaskar R, Zo SM, Narayanan KB, Purohit SD, Gupta MK, Han SS (2023) Recent development of protein-based biopolymers in food packaging applications: a review. Polym Test 124:108097. https://doi.org/10.1016/j.polymertesting.2023.108097
15. Bhatnagar P, Gururani P, Parveen A, Gautam P, Joshi NC, Tomar MS, Nanda M, Vlaskin MS, Kumar V (2024) Algae: a promising and sustainable protein-rich food ingredient for bakery and dairy products. Food Chem 441:138322. https://doi.org/10.1016/j.foodchem.2023.138322
16. Blanco-Vieites M, Casado V, Battez H, Rodríguez E (2023) Culturing Arthrospira maxima in mining wastewater: pilot-scale culturing and biomass valorisation into C-phycocyanin and crude lipid extract. Environ Technol Innov 29:102978. https://doi.org/10.1016/j.eti.2022.102978
17. Bourgougnon N, Burlot AS, Jacquin AG (2021) Algae for global sustainability? In: Jacquot JP (Ed), Advances in botanical research, past, current and future topics. Acad. Press, pp 145–212. https://doi.org/10.1016/bs.abr.2021.01.003
18. Braga ARC, Nunes MC, Raymundo A (2023) The experimental development of emulsions enriched and stabilized by recovering matter from Spirulina biomass: valorization of residue into a sustainable protein source. Molecules 28:6179. https://doi.org/10.3390/molecules28176179
19. Bürck M, Fratelli C, Assis M, Braga ARC (2024) Naturally colored ice creams enriched with c-phycocyanin and spirulina residual biomass: development of a fermented, antioxidant, tasty and stable food product. Fermentation 10(6):304. https://doi.org/10.3390/fermentation10060304
20. Cabrol MB, Glišić M, Baltić M, Jovanović D, Silađi Č, Simunović S, Tomašević I, Raymundo A (2023) White and honey Chlorella vulgaris: sustainable ingredients with the potential to improve nutritional value of pork frankfurters without compromising quality. Meat Sci 198:109123. https://doi.org/10.1016/j.meatsci.2023.109123
21. Canelli G, Tarnutzer C, Carpine R, Neutsch L, Bolten CJ, Dionisi F, Mathys A (2020) Biochemical and nutritional evaluation of Chlorella and Auxenochlorella biomasses relevant for food application. Front Nutr 7:565996. https://doi.org/10.3389/fnut.2020.565996

22. Caporgno MP, Mathys A (2018) Trends in microalgae incorporation into innovative food products with potential health benefits. Front Nutr 5:58. https://doi.org/10.3389/fnut.2018.00058
23. Carpintero M, Marcet I, Rendueles M, Díaz M (2023) Algae as an additive to improve the functional and mechanical properties of protein and polysaccharide-based films and coatings. A review of recent studies. Food Packag Shelf Life 38:101128. https://doi.org/10.1016/j.fpsl.2023.101128
24. Carvalho LF, Moreira JB, Oliveira MS, Costa JAV (2017) Novel food supplements formulated with Spirulina to meet athletes' needs. Braz Arch Biol Technol 60:e17160656. https://doi.org/10.1590/1678-4324-2017160656
25. Çelekli A, Özbal B, Bozkurt H (2024) Challenges in functional food products with the incorporation of some microalgae. Foods 13:725. https://doi.org/10.3390/foods13050725
26. Chen Y, You L, Sun-Waterhouse D (2024) Effects of processing on the physicochemical characteristics and health benefits of algae products: trade-offs among food carbon footprint, nutrient profiles, health properties, and consumer acceptance. Trends Food Sci Technol 147:104375. https://doi.org/10.1016/j.tifs.2024.104375
27. Costa JAV, Bezerra PQM, Moreira JB, Molina AN, Morais MG (2023) Biogas from microalgae: production approaches and strategies for process optimization. In: Kumar S, Kumar N, Shahid-ul-Islam (Eds), Role of microbes in industrial products and processes. Wiley. https://doi.org/10.1002/9781119901198.ch12
28. Costa JAV, Silveira JT, Cruz CG, Moreira JB, Morais MG (2022) Bioconversion of agro-wastes by microalgae for application as biofertilizers. In: Molina G, Sharma M, Benhida R, Gupta VK, Kuhad RC (eds), Microbial bioprocessing of agri-food wastes. Taylor & Francis, pp 321–344. https://doi.org/10.1201/9781003128984-11.
29. Choi D, Bedale W, Chetty S, Yu J-H (2024) Comprehensive review of clean-label antimicrobials used in dairy products. Compr Rev Food Sci Food Saf 23:13263. https://doi.org/10.1111/1541-4337.13263
30. Covacevich L, Aguilera JM, Moreno MC, Brossard N, Osorio F (2024) Rheological and tribological properties of seaweed powders as thickeners for liquid foods. Food Hydrocoll 154:110116. https://doi.org/10.1016/j.foodhyd.2024.110116
31. Da Silva SP, Valle AF, Perrone D (2021) Microencapsulated Spirulina maxima biomass as an ingredient for the production of nutritionally enriched and sensorially well-accepted vegan biscuits. LWT 142:110997. https://doi.org/10.1016/j.lwt.2021.110997
32. Dagnino-Leone J, Figueroa CP, Castañeda ML, Youlton AD, Vallejos-Almirall A, Agurto-Muñoz A, Pérez JP, Agurto-Muñoz C (2022) Phycobiliproteins: structural aspects, functional characteristics, and biotechnological perspectives. Comput Struct Biotechnol J 20:1506–1527. https://doi.org/10.1016/j.csbj.2022.02.016
33. Demarco M, Moraes JO, Matos AP, Derner RB, Neves FF, Tribuzi G (2022) Digestibility, bioaccessibility and bioactivity of compounds from algae. Trends in Food Sci Technol 121:114–112. https://doi.org/10.1016/j.tifs.2022.02.004
34. Diankristanti PA, Hei Ernest Ho N, Chen JH, Nagarajan D, Chen CY, Hsieh YM, Ng IS, Chang JS (2024) Unlocking the potential of microalgae as sustainable bioresources from up to downstream processing: a critical review. Chem Eng J 488:151124. https://doi.org/10.1016/j.cej.2024.151124
35. El-Baz FK, Abdo SM, Hussein AMS (2017) Microalgae Dunaliella salina for use as food supplement to improve pasta quality. Int J Pharm Sci Rev Res 46(2):45–51
36. Espinosa-Ramírez, J., Mondragón-Portocarrero, A.C., Rodríguez, J.A., Lorenzo, J.M., Santos, E.M., 2023. Algae as a potential source of protein meat alternatives. Front. Nutr. 10. https://doi.org/10.3389/fnut.2023.1254300.
37. Falcão RL, Pinheiro V, Ribeiro C, Sousa I, Raymundo A, Nunes MC (2023) Nutritional improvement of fresh cheese with microalga Chlorella vulgaris: impact on composition, structure and sensory acceptance. Food Technol Biotechnol 61(2):259–270. https://doi.org/10.17113/ftb.61.02.23.7851

38. Fernandes AC, Morais C, Franchini B, Pereira B, Pinho O, Cunha LM (2024) Clean-label products: factors affecting liking and acceptability by Portuguese older adults. Appetite 197:107307. https://doi.org/10.1016/j.appet.2024.107307
39. Fradinho P, Niccolai A, Soares R, Rodolfi L, Biondi N, Tredici MR, Sousa I, Raymundo A (2020) Effect of Arthrospira platensis (spirulina) incorporation on the rheological and bioactive properties of gluten-free fresh pasta. Algal Res 45:101743. https://doi.org/10.1016/j.algal.2019.101743
40. Fu Y, Chen T, Chen SY, Liu B, Sun P, Sun H, Chen F (2021) The potentials and challenges of using microalgae as an ingredient to produce meat analogues. Trends Food Sci Technol 112:188–200. https://doi.org/10.1016/j.tifs.2021.03.050
41. Ganesan AR, Shanmugam M (2020) Isolation of phycoerythrin from Kappaphycus alvarezii: a potential natural colourant in ice cream. J Appl Phycol 2020(32):4221–4233. https://doi.org/10.1007/s10811-020-02214-0
42. Gantar M, Svirčev Z (2008) Microalgae and cyanobacteria: food for thought. J Phycol 44:260–268. https://doi.org/10.1111/j.1529-8817.2008.00469.x
43. Gao K, Beardall J (2022) Using macroalgae to address UN sustainable development goals through CO_2 remediation and improvement of the aquaculture environment. Appl Phycol 3:360–367. https://doi.org/10.1080/26388081.2022.2025617
44. García AB, Longo E, Murillo MC, Bermejo R (2021) Using a B-phycoerythrin extract as a natural colorant: application in milk-based products. Molecules 26:297. https://doi.org/10.3390/molecules26020297
45. Ghosh S, Sarkar T, Pati S, Kari ZA, Edinur HA, Chakraborty R (2022) Novel bioactive compounds from marine sources as a tool for functional food development. Front Mar Sci 9:832957. https://doi.org/10.3389/fmars.2022.832957
46. Gisriel CJ, Elias E, Shen G, Soulier NT, Brudvig GW, Croce R, Bryant DA (2023) Structural comparison of allophycocyanin variants reveals the molecular basis for their spectral differences. Photosynth Res. https://doi.org/10.1007/s11120-023-01048-4
47. Gonçalves J, Freitas J, Fernandes I, Silva P (2023) Microalgae as Biofertilizers: a sustainable way to improve soil fertility and plant growth. Sustainability 15(16):12413. https://doi.org/10.3390/su151612413
48. Gordalina M, Pinheiro HM, Mateus M, Fonseca MMR, Cesário MT (2021) Macroalgae as protein sources a review on protein bioactivity, extraction, purification and characterization. Appl Sci 11(17):7969. https://doi.org/10.3390/app11177969
49. Grasso S, Estévez M, Lorenzo JM, Pateiro M, Ponnampalam EN (2024) The utilisation of agricultural by-products in processed meat products: effects on physicochemical, nutritional and sensory quality–Invited Review. Meat Sci 211:109451. https://doi.org/10.1016/j.meatsci.2024.109451
50. Hernández H, Nunes MC, Prista C, Raymundo A (2022) Innovative and healthier dairy products through the addition of microalgae: a review. Foods 11:755. https://doi.org/10.3390/foods11050755
51. Hsu JL, Sung C-C, Tseng J-T (2023) Willingness-to-pay for ready-to-eat clean label food products at convenient stores. Future Foods 7:100237. https://doi.org/10.1016/j.fufo.2023.100237
52. Inguglia ES, Song Z, Kerry JP, O'Sullivan MG, Hamill RM (2023) Addressing clean label trends in commercial meat processing: strategies, challenges and insights from consumer perspectives. Foods 12:2062. https://doi.org/10.3390/foods12102062
53. Innova Market Insights (2024) Top food trends. Accessed 7 Jul 2024. https://www.innovamarketinsights.com/trends/top-food-trends-2024/
54. Jehn FU, Dingal FJ, Mill A, Harrison C, Ilin E, Roleda MY, James SC, Denkenberger D (2024) Seaweed as a resilient food solution after a nuclear war. Earth's Future 12:e2023EF003710. https://doi.org/10.1029/2023EF003710
55. Ji L, Qiu S, Wang Z, Zhao C, Tang B, Gao Z, Fan J (2023) Phycobiliproteins from algae: current updates in sustainable production and applications in food and health. Food Res Int 167:11273. https://doi.org/10.1016/j.foodres.2023.112737

56. Kajzer M, Diowksz A (2021) The clean label concept: novel approaches in gluten-free breadmaking. Appl Sci 11:6129. https://doi.org/10.3390/app11136129
57. Khemiri S, Nunes C, Bessa RJB, Alves SP, Smaali I, Raymundo A (2021) Technological feasibility of couscous-algae-supplemented formulae: process description, nutritional properties and in vitro digestibility. Foods 10:3159. https://doi.org/10.3390/foods10123159
58. Kim HW, Choi JH, Choi YS, Han DJ, Kim HY, Lee MA, Kim SY, Kim CJ (2010) Effects of sea tangle (Lamina japonica) powder on quality characteristics of breakfast sausages. Korean J. Food Sc. Anim. Resour. 30:55–61. https://doi.org/10.5851/kosfa.2010.30.1.55
59. Kumar S, Sahoo D, Levine I (2015) Assessment of nutritional value in a brown seaweed Sargassum wightii and their seasonal variations. Algal Res 9:117–125. https://doi.org/10.1016/j.algal.2015.02.024
60. Lafarga T, Fernandez-Sevilla JM, Gonzalez-Lopez C, Acien-Fernandez FG (2020) Spirulina for the food and functional food industries. Int Food Res 137:109356. https://doi.org/10.1016/j.foodres.2020.109356
61. Lafarga T, Rodríguez-Bermúdez R, Morillas-España A, Villaró S, García-Vaquero M, Morán L, Sánchez-Zurano A, González-López CV, Acién-Fernández FG (2021) Consumer knowledge and attitudes towards microalgae as food: the case of Spain. Algal Res 54:102174. https://doi.org/10.1016/j.algal.2020.102174
62. Letras P, Oliveira S, Varela J, Nunes MC, Raymundo A (2022) 3D printed gluten-free cereal snack with incorporation of Spirulina (Arthrospira platensis) and/or Chlorella vulgaris. Algal Res 68:102863. https://doi.org/10.1016/j.algal.2022.102863
63. Li Y, Li X, Liang ZP, Chang XY, Li FT, Wang XQ, Lian XJ (2022) Progress of microencapsulated phycocyanin in food and pharma industries: a review. Molecules 27:5854. https://doi.org/10.3390/molecules27185854
64. Lopez-Lopez I, Cofrades S, Ruiz-Capillas C, Jimenez-Colmenero F (2009) Design and nutritional properties of potential functional frankfurters based on lipid formulation, added seaweed and low salt content. Meat Sci 83:255–262. https://doi.org/10.1016/j.meatsci.2009.05.014
65. Lucas BF, Rosa APC, Carvalho LF, Morais MG, Santos TD, Costa JAV (2019) Snack bars enriched with Spirulina for schoolchildren nutrition. Food Sci Technol 40:146–152. https://doi.org/10.1590/fst.06719
66. Mahmoud N, Ferreira J, Raymundo A, Nunes MC (2024) Enhancing the protein, mineral content, and bioactivity of wheat bread through the utilisation of microalgal biomass: a comparative study of Chlorella vulgaris, Phaeodactylum tricornutum, and Tetraselmis chuii. Appl Sci 14(6):2483. https://doi.org/10.3390/app14062483
67. Mamatha BS, Namitha KK, Senthil A, Smitha J, Ravishankar GA (2007) Studies on use of Enteromorpha in snack food. Food Chem 101(4):1707–1713. https://doi.org/10.1016/j.foodchem.2006.04.032
68. Manzoor MF, Afraz MT, Yilmaz BB, Adil M, Arshad N, Goksen G, Ali M, Zeng X-A (2024) Recent progress in natural seaweed pigments: green extraction, health-promoting activities, techno-functional properties and role in intelligent food packaging. J. Agric. Food Res 15:100991. https://doi.org/10.1016/j.jafr.2024.100991
69. Maruyama S, Streletskaya NA, Lim J (2021) Clean label: Why this ingredient but not that one? Food Qual Prefer 87:104062. https://doi.org/10.1016/j.foodqual.2020.104062
70. Matheus J, Alegria MJ, Nunes MC, Raymundo A (2024) Algae-boosted chickpea hummus: improving nutrition and texture with seaweeds and microalgae. Foods 13:2178. https://doi.org/10.3390/foods13142178
71. Matos AP, Novelli E, Tribuzi G (2022) Use of algae as food ingredient: sensory acceptance and commercial products. Front. Food. Sci. Technol. 2:9898012. https://doi.org/10.3389/frfst.2022.989801
72. Matos J, Afonso C, Cardoso C, Serralheiro ML, Bandarra NM (2021) Yogurt enriched with Isochrysis galbana: an innovative functional food. Foods 10:1458. https://doi.org/10.3390/foods10071458
73. Mazinani S, Fadaei V, Khosravi-Darani K (2016) Impact of Spirulina Platensis on Physicochemical properties and viability of Lactobacillus acidophilus of probiotic UF feta cheese. J Food Process Preserv 2016(40):1318–1324. https://doi.org/10.1111/jfpp.12717

74. Mendes MC, Navalho S, Ferreira A, Paulino C, Figueiredo D, Silva D, Gao F, Gama F, Bombo G, Jacinto R, Aveiro SS, Schulze PSC, Gonçalves AT, Pereira H, Gouveia L, Patarra RF, Abreu MH, Silva JL, Navalho J, Varela JCS, Speranza LG (2022) Algae as food in Europe: an overview of species diversity and their application. Foods 11:1871. https://doi.org/10.3390/foods11131871
75. Mohammadi-Gouraji E, Soleimanian-Zad S, Ghiaci M (2019) Phycocyanin-enriched yogurt and its antibacterial and physicochemical properties during 21 days of storage. LWT 102:230–236. https://doi.org/10.1016/j.lwt.2018.09.057
76. Montevecchi G, Santunione G, Licciardello F, Köker O, Masino F, Antonelli A (2022) Enrichment of wheat flour with Spirulina. Evaluation of thermal damage to essential amino acids during bread preparation. Food Res Int 157:111357. https://doi.org/10.1016/j.foodres.2022.111357
77. Moreira JB, Carvalho LF, Santos TD, Oliveira MS, Costa JAV (2018) Microalgal biotechnology for development of energy gel in sports. Int Food Res J 25:1942–1947
78. Moreira JB, Kuntzler SG, Terra ALM, Costa JAV, Morais MG (2020) Electrospun nanofibers: fundamentals, food packaging technology, and safety. In: Rangappa SM, Parameswaranpillai J, Thiagamani SMK, Senthilkumar Krishnasamy S, Siengchin S (eds), Food packaging advanced materials, technologies, and innovations. Taylor and Francis, London, pp 223–254
79. Moreira JB, Lim L-T, Zavareze ER, Dias ARG, Costa JAV, Morais MG (2018) Microalgae protein heating in acid/basic solution for nanofibers production by free surface electrospinning. J Food Eng 230:49–54. https://doi.org/10.1016/j.jfoodeng.2018.02.016
80. Moreira JB, Lim L-T, Zavareze ER, Dias ARG, Costa JAV, Morais MG (2019) Antioxidant ultrafine fibers developed with microalga compounds using a free surface electrospinning. Food Hydrocoll 93:131–136. https://doi.org/10.1016/j.foodhyd.2019.02.015
81. Moreira JB, Lucas BF, Santos TD, Duarte JH, Costa JAV, Morais MG (2024) Microalgal biotechnology: From cultivation to incorporation in a range of foods. In: Sangeetha J, Thangadurai D (eds) Algal Farming Systems: From Production to Application for a Sustainable Future. Taylor and Francis, London, pp 151–196
82. Moreira JB, Santos TD, Vaz BS, Terra ALM, Costa JAV, Morais MG (2023) Preparation of the nanocomposite films for the food packaging applications. In: Muthukumar C, Thiagamani SMK, Krishnasamy S, Parameswaranpillai J, Siengchin S (eds) Biocomposites for industrial applications. Woodhead Publishing, Elsevier, pp 313–334
83. Moreira JB, Terra ALM, Costa JAV, Morais MG (2018) Development of pH indicator from PLA/PEO ultrafine fibers containing pigment of microalgae origin. Int J Biol Macromol 118:1855–1862. https://doi.org/10.1016/j.ijbiomac.2018.07.028
84. Mosibo O, Ferrentino G, Udenigwe C (2024) Microalgae proteins as sustainable ingredients in novel foods: recent developments and challenges. Foods 13:733. https://www.mdpi.com/2304-8158/13/5/733
85. Mouritsen OG, Cornish ML, Critchley AT, Pérez-Lloréns JL (2024) History of seaweeds as a food. In: Hefft DI, Adetunji CO (eds), Applications of seaweeds in food and nutrition. Elsevier, pp 1–17. https://doi.org/10.1016/B978-0-323-91803-9.00002-0
86. Mouritsen OG, Rhatigan P, Cornish ML, Critchley AT, Pérez-Lloréns JL (2021) Saved by seaweeds: phyconomic contributions in times of crises. J Appl Phycol 33:443–458. https://doi.org/10.1007/s10811-020-02256-4
87. Muys M, Sui Y, Schwaiger B, Lesueur S, Vandenheuvel D, Vermeir P, Vlaeminck SE (2019) High variability in nutritional value and safety of commercially available Chlorella and Spirulina biomass indicates the need for smart production strategies. Bioresour Technol 275:247–257. https://doi.org/10.1016/j.biortech.2018.12.059
88. Naik B, Mishra R, Kumar V, Mishra S, Gupta U, Rustagi S, Gupta AK, Preet MS, Bhatt SC, Rizwanuddin S (2024) Micro-algae: revolutionizing food production for a healthy and sustainable future. J. Agric. Food Res. 15:100939. https://doi.org/10.1016/j.jafr.2023.100939
89. Niccolai A, Bažec K, Rodolfi L, Biondi N, Zlatić E, Jamnik P, Tredici MR (2020) Lactic acid fermentation of Arthrospira platensis (Spirulina) in a vegetal soybean drink for developing new functional lactose-free beverages. Front Microbiol 26:560684. https://doi.org/10.3389/fmicb.2020.560684

90. Noguerol AT, Pagán MJ, García-Segovia P, Varela P (2021) Green or clean? Perception of clean label plant-based products by omnivorous, vegan, vegetarian and flexitarian consumers. Food Res Int 149:110652. https://doi.org/10.1016/j.foodres.2021.110652
91. Nova P, Martins AP, Teixeira C, Abreu H, Silva JG, Silva AM, Freitas AC, Gomes AM (2020) Foods with microalgae and seaweeds fostering consumers health: a review on scientific and market innovations. J Appl Phycol 32:1789–1802. https://doi.org/10.1007/s10811-020-021 29-w
92. Nowicka-Krawczyk P, Mühlsteinová R, Hauer T (2019) Detailed characterization of the Arthrospira type species separating commercially grown taxa into the new genus Limnospira (Cyanobacteria). Sci Rep 9:694. https://doi.org/10.1038/s41598-018-36831-0
93. Nunes MC, Ferreira J, Raymundo A (2023) Volatile fingerprint impact on the sensory properties of microalgae and development of mitigation strategies. Current Op. Food Sci. 51:101040. https://doi.org/10.1016/j.cofs.2023.101040
94. O'Brien R, Hayes M, Sheldrake G, Tiwari B, Walsh P (2022) Macroalgal proteins: a review. Foods 11:571. https://doi.org/10.3390/foods11040571
95. Olabi AG, Shehata N, Sayed ET, Rodriguez C, Anyanwu RC, Russell C, Abdelkareem MA (2023) Role of microalgae in achieving sustainable development goals and circular economy. Sci Total Environ 854:158689. https://doi.org/10.1016/j.scitotenv.2022.158689
96. Oliveira APF, Bragotto APA (2022) Microalgae-based products: food and public health. Future Foods 6:100157. https://doi.org/10.1016/j.fufo.2022.100157
97. Oliveira S, Sousa I, Raymundo A (2022) Printability evaluation of Chlorella vulgaris snacks. Algal Res 68:102879. https://doi.org/10.1016/j.algal.2022.102879
98. Parmar P, Kumar R, Neha Y, Srivatsan V (2023) Microalgae as next generation plant growth additives: functions, applications, challenges and circular bioeconomy-based solutions. Front Plant Sci 14:1073546. https://doi.org/10.3389/fpls.2023.1073546
99. Pereira RN, Rodrigues R, Avelar Z, Leite AC, Leal R, Pereira RS, Vicente A (2024) Electrical fields in the processing of protein-based foods. Foods 13:577. https://doi.org/10.3390/foods1 3040577
100. Pereira T, Costa S, Barroso S, Teixeira P, Mendes S, Gil MM (2024) Development and optimization of high-protein and low-saturated fat bread formulations enriched with lupin and microalgae. LWT 191:115612. https://doi.org/10.1016/j.lwt.2023.115612
101. Prabhasankar P, Ganesan P, Bhaskar N, Hirose A, Stephen N, Gowda LR, Hosokawa M, Miyashita K (2009) Edible Japanese seaweed, wakame (Undaria pinnatifida) as an ingredient in pasta: chemical, functional and structural evaluation. Food Chem 115(2):501–508. https://doi.org/10.1016/j.foodchem.2008.12.047
102. Priyanka M, Kempanna C, Narasimha M (2013) Quality characteristics of yoghurt enriched with Spirulina powder. Mysore J. Agric. Sci. 2013(47):354–359
103. Qazi MW, de Sousa IG, Nunes MC, Raymundo A (2022) Improving the nutritional, structural, and sensory properties of gluten-free bread with different species of microalgae. Foods 11:397. https://doi.org/10.3390/foods11030397
104. Rahimah S, Sonjaya MF, Andoyo R, Satya A, Nurhasanah S, Chrismadha T (2024) Physical and organoleptic Characteristic of bread substituted with Spirulina platensis. In: Lubis MAR, Lee SH, Mardawati E, Rahimah S, Antov P, Andoyo R, Krišťák Ľ, Nurhadi B (eds), Biomass conversion and sustainable biorefinery: towards circular bioeconomy. Springer Nature, Singapore, pp 295–306. https://doi.org/10.1007/978-981-99-7769-7_14
105. Rekha VS, Sankar K, Rajaram S, Karuppiah P, Dawoud TMS, Syed A, Elgorban A (2024) Unveiling the impact of additives on structural integrity, thermal and color stability of C phycocyanin–Agar hydrocolloid. Food Chem 448:139000. https://doi.org/10.1016/j.foodchem.2024.139000
106. Rengasamy KR, Mahomoodally MF, Aumeeruddy MZ, Zengin G, Xiao J, Kim DH (2020) Bioactive compounds in seaweeds: an overview of their biological properties and safety. Food Chem Toxicol 135:111013. https://doi.org/10.1016/j.fct.2019.111013
107. Ribeiro AR, Madeira T, Botelho G, Martins D, Ferreira RM, Silva AMS, Cardoso SM, Costa R (2022) Brown algae Fucus vesiculosus in pasta: effects on textural quality, cooking properties, and sensorial traits. Foods 11:1561. https://doi.org/10.3390/foods11111561

108. Rodrigues EF, Vendruscolo LP, Bonfante K, Reinehr CO, Colla E, Colla LM (2020) Phycocyanin as substitute for texture ingredients in ice creams. Br Food J 122:693–707. https://doi.org/10.1108/BFJ-07-2019-0553
109. Sánchez-Laso J, Espada JJ, Rodríguez R, Vicente G, Bautista LF (2023) Novel biorefinery approach for phycocyanin extraction and purification and biocrude production from Arthrospira platensis. Ind Eng Chem Res 62:5190–5198. https://doi.org/10.1021/acs.iecr.2c03683
110. Sangian M, Soltani M, Hanifi H, Abdali N (2022) Investigation of the effect of phycocyanin extracted from spirulina platensis and persimmon powder on physicochemical and sensory characteristics of yogurt. Egypt J Vet Sci 53:75–86. https://doi.org/10.21608/ejvs.2021.95209.1293
111. Santos TD, Freitas BCB, Moreira JB, Zanfonato k, Costa JAV (2016) Development of powdered food with the addition of Spirulina for food supplementation of the elderly population. Innov Food Sc Emerg Technol 37:216–220. https://doi.org/10.1016/j.ifset.2016.07.016.
112. Ścieszka S, Klewicka E (2019) Algae in food: a general review. Crit Rev Food Sci Nutr 59:3538–3547. https://doi.org/10.1080/10408398.2018.1496319
113. Senthil A, Mamatha B, Vishwanath P, Bhat K, Ravishankar G (2011) Studies on development and storage stability of instant spice adjunct mix from seaweed (Eucheuma). J Food Sci Tech 48(6):712–717. https://doi.org/10.1007/s13197-010-0165-3
114. Severo IA, Lira GS, Ambati RR, Gokare RA, Vargas JVC, Ordonez J, Mariano AB (2024) Disruptive potential of microalgae proteins: shaping the future of the food industry. Future Foods 9:100318. https://doi.org/10.1016/j.fufo.2024.100318
115. Soto-Sierra L, Stoykova P, Nikolov ZL (2018) Extraction and fractionation of microalgae-based protein products. Algal Res 36:175–192. https://doi.org/10.1016/j.algal.2018.10.023
116. Stirk WA, Staden JV (2022) Bioprospecting for bioactive compounds in microalgae: antimicrobial compounds. Biotechnol Adv 59:107977. https://doi.org/10.1016/j.biotechadv.2022.107977
117. Sudhakar MP, Kumar BR, Mathimani T, Arunkumar K (2019) A review on bioenergy and bioactive compounds from microalgae and macroalgae-sustainable energy perspective. J Clean Prod 228:1320–1333. https://doi.org/10.1016/j.jclepro.2019.04.287
118. Suparmaniam U, Lam MK, Uemura Y, Lim JW, Lee KT, Shuit SH (2019) Insights into the microalgae cultivation technology and harvesting process for biofuels production: a review. Renew. Sust. Energ. Rev. 115:109361. https://doi.org/10.1016/j.rser.2019.109361
119. Surendhiran D, Li C, Cui H, Lin L (2021) Marine algae as efficacious bioresources housing antimicrobial compounds for preserving foods–a review. Int J Food Microbiol 358:109416. https://doi.org/10.1016/j.ijfoodmicro.2021.109416
120. Tan HT, Yusoff FM, Khaw YS, Ahmad SA, Shaharuddin NA (2021) Uncovering research trends of phycobiliproteins using bibliometric approach. Plants 10:2358. https://doi.org/10.3390/plants10112358
121. Terra ALM, Kuntzler SG, Costa JAV, Morais MG, Moreira JB (2023) Electrospun nanofibers based on microalgae pigments and their applications in intelligent food packaging. In: Muthukumar C, Krishnasamy S, Thiagamani SMK, Shanmugam M (eds) Electrospun nanofibres. Taylor and Francis, London, pp 236–249
122. Terra ALM, Moreira JB, Costa JAV, Morais MG (2021) Development of pH indicators from nanofibers containing microalgal pigment for monitoring of food quality. Food Biosci 44:101387. https://doi.org/10.1016/j.fbio.2021.101387
123. Thevarajah B, Piyatilleke S, Nimarshana PHV, Koushalya S, Malik A, Ariyadasa TU (2024) Exploring effective light spectral conversion techniques for enhanced production of Spirulina-derived blue pigment protein, c-phycocyanin. Bioresour Technol 399:130612. https://doi.org/10.1016/j.biortech.2024.130612
124. Thiviya P, Gamage A, Gama-Arachchige NS, Merah O, Madhujith T (2022) Seaweeds as a source of functional proteins. Phycology 2(2):216–243. https://doi.org/10.3390/phycology2020012

125. Torres-Tiji Y, Fields FJ, Mayfield SP (2020) Microalgae as a future food source. Biotechnol Adv 41:107536. https://doi.org/10.1016/j.biotechadv.2020.107536
126. Ueda JM, Pedrosa MC, Heleno SA, Carocho M, Ferreira ICFR, Barros L (2022) Food additives from fruit and vegetable by-products and bio-residues: a comprehensive review focused on sustainability. Sustainability 14:5212. https://doi.org/10.3390/su14095212
127. United Nations Department of Economic and Social Affairs, Sustainable Division, 2023. THE 17 GOALS (2024). URL https://sdgs.un.org/goals
128. Uribe-Wandurraga ZN, Igual M, Reino-Moyón J, García-Segovia P, Martínez-Monzó J (2021) Effect of microalgae (Arthrospira platensis and Chlorella vulgaris) addition on 3D printed cookies. Food Biophys 16:27–39. https://doi.org/10.1007/s11483-020-09642-y
129. Ververis E, Ackerl R, Azzollini D, Colombo PA, Sesmaisons A, Dumas C, Fernandez-Dumont A, Costa LF, Germini A, Goumperis T, Kouloura E, Matijevic L, Precup G, Roldan-Torres R, Rossi A, Svejstil R, Turla E, Gelbmann W (2020) Novel foods in the European Union: scientific requirements and challenges of the risk assessment process by the European food safety authority. Food Res Int 137:109515. https://doi.org/10.1016/J.FOODRES.2020.109515
130. Vignesh K, Atchaya R, Pavan Kumar Rao G, Shraddha Balan Jaffar Ali BM, Emerson Andrade Sales, Arun Prasath R, Donato AG (2023). A cascaded biorefinery for the sustainable valorisation valorization of Arthrospira maxima biomass: a circular bioeconomy approach. Bioresour Technol Rep 23:101510. https://doi.org/10.1016/j.biteb.2023.101510
131. Wang M, Yin Z, Sun W, Zhong Q, Zhang Y, Zeng M (2023) Microalgae play a structuring role in food: Effect of spirulina platensis on the rheological, gelling characteristics, and mechanical properties of soy protein isolate hydrogel. Food Hydrocoll 136:108244. https://doi.org/10.1016/j.foodhyd.2022.108244
132. Wang T, Jonsdottir R, Kristinsson HG, Thorkelsson G, Jacobsen C, Hamaguchi PY, Olafsdottir G (2010) Inhibition of haemoglobin-mediated lipid oxidation in washed cod muscle and cod protein isolates by Fucus vesiculosus extract and fractions. Food Chem 123:321–330. https://doi.org/10.1016/j.foodchem.2010.04.038
133. WHO (2007) World Health Organization Protein and amino acid requirements in human nutrition: report of a joint WHO/FAO/UNU expert consultation. Geneva. Accessed 20 Mar 2024. https://apps.who.int/iris/handle/10665/43411
134. Wolkers H, Barbosa M, Kleinegris DMM, Bosma R, Wijffels RH (2011) Microalgae: the green gold of the future? In: Harmsen P (ed) Large-scale Sustainable Cultivation of Microalgae for the Production of Bulk Commodities. Propress, Wageningen, pp 1–34
135. Wu JY, Tso R, Teo HS, Haldar S (2023) The utility of algae as sources of high value nutritional ingredients, particularly for alternative/complementary proteins to improve human health. Front Nutr 10. https://doi.org/10.3389/fnut.2023.1277343.
136. Yamaguchi SKF, Moreira JB, Costa JAV, de Souza CK, Bertoli SL, Carvalho LFD (2019) Evaluation of adding Spirulina to freeze-dried yogurts before fermentation and after freeze-drying. Ind Biotechnol 15:89–94. https://doi.org/10.1089/ind.2018.0030
137. Yang S, Wang Y, Wang J, Cheng K, Liu J, He J, Zhang Y, Mou H, Sun H (2024) Microalgal protein for sustainable and nutritious foods: a joint analysis of environmental impacts, health benefits and consumer's acceptance. Trends Food Sci Technol 104278. https://doi.org/10.1016/j.tifs.2023.104278
138. Young M, Paul N, Birch D, Swanepoel L (2022) Factors influencing the consumption of seaweed amongst young adults. Foods 11:3052. https://doi.org/10.3390/foods11193052
139. Zhang Q, Fu B, Chen Q, Lu J, Zhu Z, Yan C, Guan F, Wang P, Fu L, Yu P (2024) Biosynthesis of the phycocyanin β-subunit in Escherichia coli BL21 and its antioxidant activity and application in the preservation of fresh-cut apples. Int J Biol Macromol 258:128951. https://doi.org/10.1016/j.ijbiomac.2023.128951
140. Zhong H, Gao X, Cheng C, Liu C, Wang Q, Han X (2020) The structural characteristics of seaweed polysaccharides and their application in gel drug delivery systems. Mar Drugs 18:658. https://doi.org/10.3390/md18120658

Microalgal Technology Challenges

Maya Margaritova Zaharieva, Fabiano Bussinela Scheifelle, Alexander Dimitrov Kroumov, and Hristo Najdenski

Abstract Microalgae technology value is well known. Many achievements of the researchers highlight their potential to be used as a renewable energy source. Hence, focus will be given on microalgae as challenging competitive sources of biofuels. Development of cost-effective microalgae technology for biofuels requires the understanding and application of the biorefinery concept by the utilization of every component of biomass, recycling harvested water and using wastewaters as a source of nutrients for the growth of microalgae. The complexity of the whole productive chain where lipids are used for biodiesel requires the application of innovative decisions and mathematical methods for process development and photobioreactor (PBR) design and scale-up. This chapter analyzes the kinetics of the transesterification process of lipids to biodiesel as well as the mathematical approach as a powerful tool for the analysis and development of any unit operation, and influence of state parameters on microalgal growth. Special attention is given to microalgal strain selection, complex mathematical analysis of closed column PBR design whose effectiveness determines the success of overall technological scheme of biomass production, lipids, and other target metabolites.

1 Introduction

In this book chapter, the microalgae's potential as a source of bioenergy precursors such as lipids metabolites. Special and complex evaluation will be made on all factors and system operation influencing maximization of biomass and products

M. M. Zaharieva · A. D. Kroumov (✉)
Department of Infectious Microbiology, The Stephan Angeloff Institute of Microbiology, Bulgarian Academy of Sciences, Sofia, Bulgaria
e-mail: adkrumov@gmail.com

F. B. Scheifelle
Universidade Tecnológica Federal Do Paraná, Toledo, Brazil

H. Najdenski
Engenharia de Bioprocessos e Biotecnologia–COEBB, Parana, Brazil

A. Khan and K. A. Alzahrani (eds.), *Microalgae Biotechnology and Sustainable Development*, Sustainable Materials and Technology,
https://doi.org/10.1007/978-981-95-2597-3_4

most particularly of interests such as a bioenergy source. Green clean energy has been actively investigated in order to satisfy the increasing worldwide energy demands [170]. Microalgal technology offers biodiesel from algal lipids as an alternative to replace fossil fuels because of many advantages which can be offered in the near future comparing with other clean energies most often used for power generation and heating purposes [48]. Major advantages of microalgae as third- and fourth-generation biofuels are their ability for growing easily by using simple nutrients in the environment such as freshwater, seawater, wastewater, food waste, etc., and most importantly to obtain higher biomass than land plants [30, 120, 125, 151, 207, 215]. The microalgae's fast growth rate and their corresponding higher photosynthetic efficiency resulted in a higher yield of bioenergy products [87] such as bioethanol [87, 160, 166, 184, 201], biodiesel [129, 145, 163], and biohydrogen [79, 107, 158]. Biodiesel is a key alternative to biofuels, and many efforts have been made to study its production systems. Microalgae species have been identified as its producers, including *Chlorella vulgaris, Nannochloropsis oceanica, Dunaliella salina, Botryococcus, Desmodesmus, Neochloris, Scenedesmus,* and *Tetraselmis*, being prospective for biodiesel production. Laboratory scale was used to prove the hypothesis and meet ASTM standards [5, 27, 64, 68, 124, 140, 222, 226]. It must be noticed that an important issue connecting with the fuel properties is to study the proportion of mixture between diesel and biodiesel from microalgae [153, 194, 208].

As many authors mentioned, the strain selection is a key to obtaining cost-effective technology of biofuel production [28]. In order to improve tremendously the lipid content in industrial microalgae strains, genetic manipulation of cell genome was of special attention to scientists in the field and as a result species such as *Chlamydomonas, C. vulgaris, Phaeodactylum tricornutum, Thalassiosira pseudonana, Galdieria sulphuraria, B. braunii,* and *D. salina* have been modified [174]. Random mutagenesis cannot compete with this approach in effectiveness [185]. Successful overall effectiveness of process operations is based on the achievements of optimal downstream methods. The genetic transformation of the cells improves their potential for synthesis of targeted metabolites [30, 72, 83, 99, 139, 195].

Crucially important is to get and systematize the knowledge about the biodiesel from microalgae lipids. Hence, we have to show the usefulness of modeling describing the chemical reactions of transesterification processes [219], to outline the advantages of using enzymatic methods of catalysis [96] and other ones without the application of chemical catalysts (KOH is the most often used) which at the end of the process remains in the glycerol diminishing its quality. It is valuable to analyze the modeling stages of the transesterification process of soybean oil from the paper of [218]. The model described the reverse mechanism of transesterification reactions in dynamics. Profiles of triglycerides (TG), diglycerides (DG), monoglycerides (MG), biodiesel, and glycerol concentrations were obtained. Model parameters were identified on the basis of different experimental data for homogeneous catalytic and non-catalytic processes published in the literature which proved the accuracy of the developed model. The authors showed nice correlation between model simulations and experimental data.

2 Model Development

The overall transesterification reactions of TG with alcohol (ROH, usually methanol) into methyl esters (biodiesel, BD) can be written as follows [60]:

$$TG + ROH \underset{K_{11}}{\overset{K_1}{\rightleftarrows}} DG + BD \tag{1}$$

$$DG + ROH \underset{K_{22}}{\overset{K_2}{\rightleftarrows}} MG + BD \tag{2}$$

$$MG + ROH \underset{K_{33}}{\overset{K_3}{\rightleftarrows}} GL + BD \tag{3}$$

The overall reaction is

$$TG + 3ROH \rightarrow GL + 3BD \tag{4}$$

Assumptions of the represented mechanism are as follows: homogeneous catalytic reversible reactions; kinetics is first order; the mass transfer and transport phenomena limitations are neglected. The kinetics constants can be represented by the Arrhenius model.

The transesterification equilibrium or forward reactions can be guided by the catalyst (if any is involved) and molar ratio Alcohol:Soybean Oil (ROH:SBO). Hence, the kinetics of the transesterification process can be written by using mass balances of the components:

$$\frac{dC_{TG}}{dt} = -K_1.C_{TG}(t).C_{ROH}(t) + K_{11}.C_{DG}(t).C_{BD}(t) \tag{5}$$

$$\frac{dC_{DG}}{dt} = K_1.C_{TG}(t).C_{ROH}(t) - K_{11}.C_{DG}(t).C_{BD}(t) - K_2.C_{DG}(t).C_{ROH}(t) + \\ + K_{22}.C_{MG}(t).C_{BD}(t) \tag{6}$$

$$\frac{dC_{MG}}{dt} = K_2.C_{DG}(t).C_{ROH}(t) - K_{22}.C_{MG}(t).C_{BD}(t) - K_3.C_{MG}(t).C_{ROH}(t) + \\ + K_{33}.C_{GL}(t).C_{BD}(t) \tag{7}$$

$$\frac{dC_{GL}}{dt} = K_3.C_{MG}(t).C_{ROH}(t) - K_{33}.C_{GL}(t).C_{BD}(t) \tag{8}$$

$$\frac{dC_{BD}}{dt} = -3.\left(\frac{dC_{TG}}{dt}\right) - 2.\left(\frac{dC_{DG}}{dt}\right) - 1.\left(\frac{dC_{MG}}{dt}\right) \quad (9)$$

$$\frac{dC_{ROH}}{dt} = -\frac{dC_{BD}}{dt} \quad (10)$$

The system (eqs.1–5) was transformed as shown in the works of [40, 44], where the concentration of each component was normalized as a mass ratio $X_{component}$ (kg component/kg ester):

$$\frac{dX_{TG}}{dt} = -K_1.X_{TG}(t).C_{ROH}(t) + \frac{PM_{TG}}{PM_{DG}}K_{11}.X_{DG}(t).C_{BD}(t) \quad (11)$$

$$\begin{aligned}\frac{dX_{DG}}{dt} &= \left(\frac{PM_{DG}}{PM_{TG}}K_1.X_{TG}(t) - K_2.X_{DG}(t)\right)C_{ROH}(t) + \\ &+ \left(\frac{PM_{DG}}{PM_{MG}}K_{22}.X_{MG}(t) - K_{11}.X_{DG}(t)\right)C_{BD}(t)\end{aligned} \quad (12)$$

$$\begin{aligned}\frac{dX_{MG}}{dt} &= \left(\frac{PM_{MG}}{PM_{DG}}K_2.X_{DG}(t) - K_3.X_{MG}(t)\right)C_{ROH}(t) + \\ &+ \left(\frac{PM_{MG}}{PM_{GL}}K_{33}.X_{GL}(t) - K_{22}.X_{MG}(t)\right)C_{BD}(t)\end{aligned} \quad (13)$$

$$\frac{dX_{GL}}{dt} = \frac{PM_{GL}}{PM_{MG}}K_3.X_{MG}(t).C_{ROH}(t) - K_{33}.X_{GL}(t).C_{BD}(t) \quad (14)$$

$$X_{BD}(t) = 1 - X_{TG}(t) - X_{DG}(t) - X_{MG}(t) \quad (15)$$

where

$$C_{ROH}(t) = \frac{n_{ROH} - 3.n_{TG}(1 - X_{TG}(t) - X_{DG}(t) - X_{MG}(t))}{(PM_{TG}n_{TG} + PM_{ROH}n_{ROH})/1000}$$

and

$$C_{BD}(t) = \frac{3.n_{TG}(1 - X_{TG}(t) - X_{DG}(t) - X_{MG}(t))}{(PM_{TG}n_{TG} + PM_{ROH}n_{ROH})/1000}.$$

where n_{TG} is the number of SBO mols; n_{ROH} is the number of alcohol mols; $PM_{component}$ represents the molecular weight of the component.

The rate constants involved in the forward (K_i) and reverse reactions (K_{ii}) are

$$K_i = K_{i,0}.e^{-\frac{Ea_i}{R.T}} \quad (16)$$

$$K_{ii} = K_{ii,0}.e^{-\frac{Ea_{ii}}{R.T}} \tag{17}$$

where Ea_i stands for energy of activation for the ith forward reaction; Ea_{ii} stands for energy of activation for reverse reaction, R is a gas constant, and T stands for reaction temperature. $K_{i,0}$ and $K_{ii,0}$ stand for initial rate constants of forward and reverse reactions; $C_{component}$ is component concentration [kg/m^3]; $X_{component}$ is mass ratio [kg component/kg ester]; K_i is ith forward kinetic rate constant [kg total/(mol ROH.h)]; K_{ii} is iith reverse kinetic rate constant [kg total/(mol BD.h)].

The authors used normalized model (Eqs. 11–15) to fit experimental data of transesterification catalytic and non-catalytic processes under various operational conditions.

The authors checked the accuracy of the developed model (Maple® software) on experimental data taken from the literature [44, 60] and strong evidence of its robustness and rightness. Details about the nonlinear parameter identification procedure can be found in our paper [218].

Results are presented in Figs. 1 and 2. In Fig. 1 are shown dynamic changes of TG, DG, MG, and BD concentrations versus time for the given reaction conditions. The solid curves are model profiles, and points represent experimental data. In Fig. 1 TG concentration continuously decreases up to zero. Changes of DG concentration are the changes that are indicative because DG increases rapidly at the begging up to its maximum and after that slowly decreases but it did not reache zero at the end. Monoglyceride concentration slowly increases during the process, and its utilization is not completed at end of the reaction time. The residual concentration of MG of about 15% has shown that the third forward trasnesterification reaction is slowest and determines the limitation conditions of the overall transesterification process. The model profiles of TG and DG have almost perfect match with experimental data. The BD profile did not match perfectly the experimental one. The same but in bigger amplitude is observed for MG profile; nevertheless the model behavior is excellent.

Fig. 1 Simulation results for transesterification of soybean oil and experimental results from work of Disakou and co-authors [44]. Reaction conditions: temperature 220 °C, molar ratio ROH/TG 27/1, alcohol methanol, and non-catalytic reaction

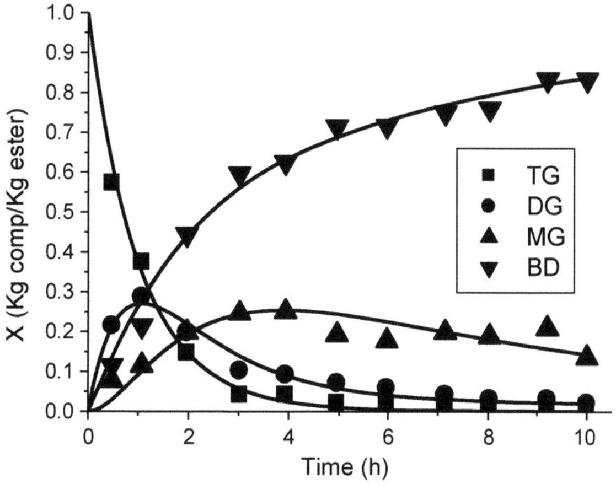

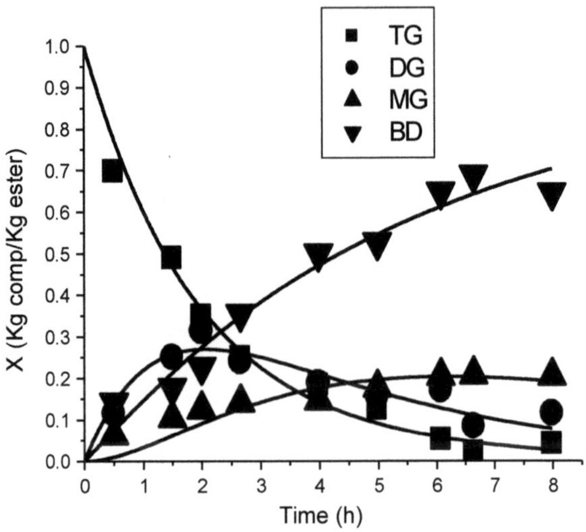

Fig. 2 Simulation and experimental results for transesterification of soybean oil. Reaction conditions: temperature 235 °C, molar ratio ROH/TG 21/1, alcohol methanol, and non-catalytic reaction [44]

Analysis of Fig. 2 on TG, DG, and MG profiles has shown that the TG concentration decreases with the same trend as in Fig. 1, but DG concentration profile differs from the one in Fig. 1 not reaching zero. The MG concentration reaches the plateau from fifth up to eighth hour. The reaction temperature of this process is higher with 15 °C than the one presented in Fig. 1, but molar ratio Methanol/TG is lower (21/1). If the final product BD is not removing from the reaction mixture, molar ratio appears to be the key parameter for maximum BD production. Therefore, BD productivity is higher for the process shown in Fig. 1.

Analysis of Fig. 4 has shown that the concentration of the catalyst plays a key role for conversion of reagents, and time is bigger than for the process shown in Fig. 3. The 10 °C difference is not enough to compensate the lower catalyst concentration. The TG, DG, MG, and BD model profiles match excellently the sets of experimental data. The conversion of TG to BD is higher than in the process shown in Fig. 3. This fact can be explained by the used alcohol. It has to be noticed that for the BD production, methanol reaches higher conversion rates than other primary alcohols.

If the base catalyst is applied, the forward reactions dominate and the process time is up to 30 h depending on state parameters and operational conditions.

A comparison between acid and alkali catalysts applied in the transesterification processes of soybean oil has shown that the alkali catalyst is preferable over the acid one. Non-catalytic transesterification process is very challenging because of the purity of produced glycerol. It must be noticed that the requirements for high pressure and high temperatures need special equipment for realization of the transesterification process. In our work [218], it was shown that modeling of kinetics of transesterification of soybean oil give crucial knowledge of how to develop an effective optimal technology of BD production. The developed new kinetics model can be extremely important for teaching under-graduate, graduate, and post-doctoral

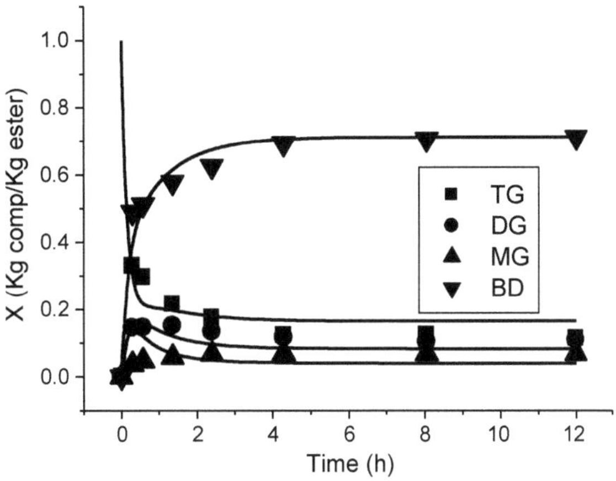

Fig. 3 Dynamic simulation results for transesterification of soybean oil and experimental results from work of [60]. Reaction conditions: reaction temperature 30 °C, molar ratio ROH/TG 3/1, ROH = butanol, and catalyst = 1% NaOBu

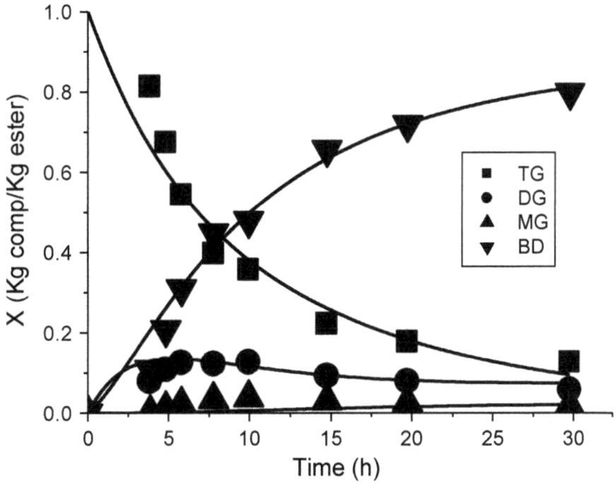

Fig. 4 Dynamic simulation results and experimental results from work of [60]. Reaction conditions: soybean oil, reaction temperature 40 °C; molar ratio ROH/TG 3/1; ROH = Methanol; catalyst = 0.5% $NaOCH_3$

students because this knowledge can be applied with some adoptions for any lipids obtained from microalgae biomass.

Challenging opportunity to make BD by using enzymatic catalyst as lipase needs knowledge about mechanisms of action of the enzyme in such a transeterification process [96]. The authors proposed the enzymatic kinetics model of vegetable oil transesterification for BD production. It is well known that enzymatic production of biodiesel is more promising than that based on chemical methods using acid or base catalysts. The immobilized enzyme lipase used as catalyst in this process was isolated from microorganisms such as *Pseudomonas fluorescens, P. cepacia, Candida antarctica.* The operational conditions in all fitted experimental data involved butanol and propanol as alcohols, molar ratio 3:1 of ROH:TG, and temperature 50 °C. The process was modeled considering three consecutive reactions where DG and MG were formed. Further, three intermediate stages involving lipase were considered.

The simulation results showed good model flexibility when applied on different sets of data taken from the literature. Therefore, using principles of analogy will be very useful to apply similar modeling procedure for analysis and transesterification process design of any lipids obtained from microalgae biomass.

Complex analysis needs to be applied not only on the final stage where lipids were extracted from microalgae biomass and used for BD production. Especially important is to analyze unit operations in upstream processes such as microalgae strain isolation and culturing in order to maximize lipid content in the cells of the chosen strain of interest. An original paper showing such an approach demonstrates interesting findings [62] in the state of the art, culturing four microalgae species by using different culturing techniques (photoautotrophic, mixotrophic, and heterotrophic conditions) and connecting them with biomass growth and lipids content in the cells. Most particularly, the authors studied and compared the microalgal growth and total lipid content of four green microalgae such as *Chlorella sp., Chlorella vulgaris CCAP211/11B, Botryococcus braunii FC124,* and *Scenedesmus obliquus R8* grown under autotrophic, heterotrophic, and mixotrophic conditions. The influence of six carbon sources (glucose, xylose, fructose, sucrose, and galactose) and different glucose concentrations and light irradiations on cell growth and total lipid content were under evaluation. The results from this original work showed that the four algae growth rates and total lipid contents were investigated in three different modes of cultivations. A well-known fact is that heterotrophic cultivation gives higher biomass than autotrophic one, and it was observed once again in this work. From the chosen carbon sources, glucose was the best one for four microalgae growth. The applied 1% glucose was optimal for biomass synthesis in three microalgae species. An exception was the strain *S. obliquus R8* (2% glucose). The *B. braunii FC124* was found to be the best lipid produced by the cells with less than 80 μmol.m^{-2}.s^{-1} of light intensity, 1% of glucose, and 20 days incubation. Lipid accumulation increased with increasing light intensity in the mixotrophic cultivation. Total lipid content was lesser in the heterotrophic mode when compared with other two cultivation techniques. Nevertheless, no significant variations in the biomass production was noted between heterotrophic and mixotrophic cultivations. The highest total lipid content was observed at 100 μmol.m^{-2}.s^{-1} for *Chlorella sp.* (22.5%) and *S. obliquus R8* (23.7%). For other two species, 80 μmol.m^{-2}.s^{-1} *C. vulgaris CCAP211/11B* (20.1%) and *B. braunii FC124* (34.9%) gave maximum lipid value. The authors claimed that these microalgae species are promising to be used for the industrial production of lipids. It must be noted that many researchers found that *B. braunii* species accumulate extremely high content of lipids under stress and different light illuminating conditions [53]. It means that all major factors influencing lipid synthesis in microalgae must be understood and systematized in catalogs. Here, step by step we briefly will follow the achievements obtained from strain selection as well as the influence of state parameters and operational conditions.

3 Selection and Cultivation of Microalgae Strains

The success of large-scale microalgae culture relies heavily on the selection of appropriate species [176]. The authors present a strategy for isolation, identification, and evaluation of microalgae strains (*Poterioochromonas malhamensis, Chlorella sp., Micractinium sp., Tetradesmus sp., and Desmodesmus sp.*) and features when culturing in PBRs with the goal to synthesizing biologically active compounds (BAC). Their methodology can be extended for overproduction of any metabolites including lipids. Especially valuable can be this work to show how the two-step cultivation process may achieve optimal lipid productivity in industrial-scale PBRs.

The knowledge is connected to green algae (*Chlorophyceae*) which have a good potential for biodiesel production [39, 73, 152]. Most often, the average lipid content of green algae is about 25.5%. Controlling the cultivation by stress factors (for example, nutrients limitations) can elevate the lipid content up to 45.7% [73]. For industrial application of biodiesel production, lipids of green algae are preferable by many reasons than lipids isolated from crops. Active research has been done for screening new microalgae strains with high lipid content. Many works illustrated the efforts in this state of the art. For example: *C. vulgaris* may accumulate lipids in the range of 11.9% to 54.2% w/w_{dw} [9, 43, 140, 229],*Chlorella sorokiniana* 44% w/w_{dw} lipid content [100, 110], *Scenedesmus incrassatulus* between 19.7% and 37.7% w/w_{dw} lipid content [55], *Tetradesmus obliquus (Scenedesmus obliquus)* 16.73%–55% w/w_{dw} lipid content [55, 121, 140, 178, 211]. The maximum lipid content of the species *Neochloris oleoabundans* and *Botryococcus braunii* can reach more than 74%. In the work of authors [230], one may see that combination of high lipid content and biomass concentration in *Chlorella* production cost can be successfully reduced.

Another important issue connects with the fatty acids (FAs) composition because they affect the biodiesel quality. Microalgae FAs consist of both saturated and unsaturated FAs, where hydrocarbon chains include from 4 to 36 carbons [37]. It can be noticed that the main saturated fatty acids (SFA), monounsaturated fatty acids (MUFA), and polyunsaturated fatty acids (PUFA) in microalgae are as follows: palmitic acid (C16:0), oleic acid (C18:1), linoleic acid (C18:2), and linolenic acid (C18:3), respectively [58]. Comparison of FA profiles of different microalgae species, in this study, showed the following: SFA and MUFA are dominant in most of the algae species. The sum of SFA and MUFA in *A. fusiformis, B. braunii, Chlamydomonas sp., C. vulgaris, Nannochloris sp., N. oleoabundans, S. incrassatulus, T. obliquus, and P. subcapitata* remarkably were more than two-thirds of the total FAs. Therefore, the MUFA can be the best choice for biodiesel production [98]. The studied microalgae synthesized big enough of C:16 and C:18 acids in particular these are elevated levels of C16:0 and C18:1, which formed major structural components in biodiesel [104]. Therefore, the selection of suitable strains for biodiesel production should take into account the FA composition in microalgae.

It must be noticed that during the process development and optimization of microalgae systems for biodiesel production, it must take into account the phenomenon that most microalgae strains achieving high biomass concentration do not present high lipid productivity. It means that the two-step optimization procedure must be used in both, high lipid content and high biomass productivity. Crucially important for the cost-effective process is to apply such an optimization approach which at the bigger or industrial scale of microalgae cultivation in huge PBRs provides successful technological solution in the market.

4 Co-Cultivation with Microorganisms

Special attention must be given when analyzing the WW treatment plants where microalgae co-exist with many other microorganisms in complex microbial communities. This evidently has a significant impact on microalgae growth and lipid production [26]. These interactions must be studied in order to get maximum benefits from microalgae and microorganisms symbiosis. The O_2 produced by the microalgae during photosynthesis is used by microorganisms, which in return produced CO_2 as a source of carbon for microalgae. Here, important is to notice that there is no competition for the carbon source because microorganisms and microalgae consume organic and inorganic carbon, respectively. These phenomena are responsible for enhanced biomass production and higher concentrations of secondary metabolites of interest, in our case lipids.

Recently, Wei et al. studied co-cultivation of mixed culture of *Chlorella vulgaris* and *Mesorhizobium sangaii* under nitrogen limiting conditions [217]. They found that the maximum biomass, lipid content, and productivity of algae in the co-cultures were 1.5, 2.2, and 3.3 times higher than those of the pure microalgae culture. The lipid productivity was 3.5-fold higher compared to control, where 50% of neutral lipid proportion was observed. The content of C18:1 and C18:2 increased because of the chosen consortium [217]. Another beneficiary technique was demonstrated such as a continuous separated two-step (CSTS) cultivation of mixed culture. Firstly, *Chlorella saccharophila* was grown in a bioreactor, while *Cellvibrio pealriver* was grown in a dialysis bag inside the bioreactor. In one moment, the inner bag was removed and the second-step cultivation of *C. saccharophila* started. The authors observed immediate increase of microalgae lipid content after removing the inner bag. The lipids reached 53.89% after 15 days cultivation, which was near 2.5-fold higher than the one-step co-cultivation. Comparing the microalgae lipid values of CSTS with one-step co-cultivation showed values such as 929.79 and 607.90 mg.L^{-1}, respectively [224]. The authors of another work have studied *Chlorella vulgaris* and *Streptomyces rosealbus* in co-cultivated scenario with the same goal to find the potential of co-culture system on lipid synthesis and microalgae growth. The co-cultured microalgae showed 33% higher biomass concentration than the control. The lipid production increased by 80%. This follows from synergistic symbiosis between *C. vulgaris* and *S. rosealbus* [102]. Arora et al. have studied the lipid biosynthesis pathway of yeast and microalgae and links between them depending on conditions and state parameters of the system [8]. According to them, the strain selection is a key parameter. The best results were found when co-culture system was formed from the genera of *Rhodotorula sp.* and *Chlorella sp.* The optimal value (1:1) of microalgae/yeast ratio was observed. Chu et al. demonstrated positive symbiosis of microalgae/fungus of mixed culture to enrich the biomass and lipid concentrations [32]. The chosen species for this study were *Chlorella protothecoides* and *Aspergillus fumigatus*. Significant increase of biomass and lipid values was observed such as 2- and 2.15-fold higher compared to the monocultures, respectively.

Finally, it is fundamental to study co-culturing microalgae bacteria, yeast fungi, and their different combinations because WW contained such consortiums from any source—industry, agriculture, and domestic WW.

5 Hormones

Recently, the interest to study application of hormones to improve lipid production in microalgae has increased. Jiang et al. pointed out that enough low concentration of DA-6 (10 − 7 M) enhances the growth of *Scenedesmus quadricauda SDEC-13* and *Chlorella ellipsoidea SDEC-11* and, remarkably, biomass productivity was 14 and 336% higher than without DA-6, respectively [78]. Besides, the higher lipid productivity occurred of about 30% and 15% than in control cultures [78]. In another study, Salama et al. demonstrated that indole-3-acetic acid (IAA, auxine) and diethyl aminoethyl hexanoate (DA-6) influenced better growth and lipid productivity of the microalgae species *Scenedesmus obliquus GU732418*, *Ourococcus multisporus GU732424,* and *Chlorella vulgaris FR751187* [171].

Supplementation of nutrient medium with 10 $mg.L^{-1}$ of indole-3-butyric acid (IBA) and indole-3-propionic acid (IPA) when culturing *Desmodesmus sp. JS07* significantly increases the lipid value of biomass. Higher lipid content was observed when added 10 $mg.L^{-1}$ benzylaminopurine (BAP). Examining the combination of IBA (10 $mg.L^{-1}$) and BAP (5 $mg.L^{-1}$) showed higher biomass concentration and lipid content in the strain highlighting a synergistic effect [180].

Another hormone, the strigolactone was used on *Monoraphidium sp. QLY-1* cultures and led to enhance lipid content and lipid productivity under 1 μM induction [188]. On the other hand, application of stress factor such as nitrogen limitation outperformed strigolactone induction [187]. Research on *Nannochloropsis oceanica CASA CC201* demonstrated that addition of 10 ppm of methyl jasmonate (MeJA) increased biomass and lipid synthesis of 22% and 49% respectively compared with control culture [209]. Further, an increase of MeJA concentration beyond 10 ppm induced a decrease of essential ω-3 fatty acid but promoted oleic acid synthesis which has to be evaluated in the frame of cost-effectiveness.

A novel growth regulator such as the p-coumaric acid was found in agricultural WW. Addition of 1 mM of it increased the lipid accumulation up to 30.7% in *Tetradesmus obliquus BPL16*. Details are published elsewhere [52].

6 Optimization Methodology

Hence, the single-parameter approach for optimization of any biotechnological process nowadays is avoided as less valuable. Hence, to improve the lipid production in microalgae requires a methodology where many parameters and their interactions are considered in the optimization procedure. Hence, it is important to show reports presenting the positive effects of combining abiotic stress factors to improve the lipid production in microalgae. Kwak et al. demonstrated the work where the combined effect of various stress conditions on lipid productivity in *Chorella, Scenedesmus, Chlamydomonas,* and *Neochloris* strains were studied [101]. Compared to the single nitrogen depletion stress, combined with high salt concentration has improved the lipid productivities of *Chlamydomonas* species up to 50% and for *Chlorella protothecoides*—30%. The lipid synthesis of *C. vulgaris* and *N. oleoabundans* were higher up to 39% and 37%, respectively under the combined condition of N-starvation and 30 °C temperature compared to the maximum achieved under the single stress. Culturing *C. zofingiensis*, the increase of lipid production up to 44% was achieved with the combined effects of temperature (30 °C) and high NaCl concentration. The optimal combined stress for *Scenedesmus sp.* was found to be under the high temperature and pH = 10. It resulted in 1.54-fold higher lipid productivity compared to one under single factor of pH = 10. The lipid content of *H. pluvialis* was increased in combination of melatonin, nitrogen deficiency, and light illumination at 150 μmol photons.m^{-2}.s^{-1}. The result was highest lipid content—42.84%—which was 1.22-fold greater than in controls [46]. In *Chlorella sp. BUM11008*, the combination of iron and potassium-phosphate limitations achieved a high increase of lipid content of biomass especially values of linolenic and arachidic acid [154].

More recently, [187] have demonstrated the combined effect of nitrogen deficiency with strigolactone (SL). They observed an increase of the lipid content and lipid productivity in *Monoraphidium sp. QLY-1.* Also, the jasmonic acid (JA) content was increased rapidly under nitrogen deficiency with or without addition of SL. Further, an exogenous JA addition was studied and led to the highest lipid content (57.12%) and productivity (322.65 mg.L^{-1}.d^{-1}) under nitrogen stress combined with SL treatment. It proves the value of JA in regulating lipid synthesis in microalgae [187]. Other work of Zhao et al. analyzed the effect of melatonin on lipid synthesis in *Monoraphidium sp. QLY-1* under the condition of high salinity [233]. The melatonin induction resulted in significantly high lipid content, which was 1.16- and 1.44-fold higher than under salinity stress alone and control, respectively. Very recently, Chen et al. study the influence of glycerol and corn steep liquor on metabolism of *Thraustochytrium sp. BM2* for robust lipid synthesis [22]. The lipid productivity was 76% and 37 mg.L^{-1}.h^{-1} at 12.5 g.L^{-1} glycerol and 10 g.L^{-1} liquor. Lipid content depends on addition of sea salt (2% w/v) and its value was enhanced up to 79%. Culturing technique in a semi-continuous mode allows to reach overall lipid productivity up to 158 mg.L^{-1}.h^{-1} [21]. Research on mixed population of an isolated co-culture microalgae-yeast (CCMY) from WW was considered as biomass and lipid producer. The consortium includes microalgae as *Scenedesmus obliquus, Scenedesmus sp.*, and yeast as *Candida pimensis*. A Box-Behnken Experimental Design (BBED) optimized the values of nutrients on the biomass and lipids synthesis. The values of three variables were changed. As a result, a maximum lipid value in CCMY was achieved when the ratio of sodium nitrate:potassium phosphate was the smallest and the potassium phosphate was lower than another key chemical agent as ferrous sulfate [193]. In another work, the same statistical Box-Behnken design was used to optimize the value of six variables on lipid productivity of *Chlorella protothecoides* [14]. The optimum was determined for the chosen state parameters, and the highest lipid productivity and CO_2 utilization were found 274.15 $mg.L^{-1}.d^{-1}$ and 273.66 $mg.L^{-1}.d^{-1}$, respectively. Therefore, the biomass and lipid content can be optimized in single- or two-step experimental designs where first is the optimization of biomass concentration and after that optimization continue using multifactor design to control metabolic synthesis of lipids.

Always the final decision about any technology is based on techno-economical consideration which determines the success in competitiveness in the market. Hence, the production costs of biomass and lipids will depend on state parameters and engineering approaches to optimize the upstream and downstream processes. The specific features of microalgae cells determined their response to the applied stress factors synthesizing different fatty acids and other metabolites. Special attention has to be given on working conditions and key components of the medium which are important for maximization of lipid production in microalgae. Below, we are going to present state of the art in the area where changes of CO_2, $NaHCO_3$, nitrogen, light, temperature, salt stress, and trace elements of the medium and addition of hormones influence lipid synthesis in a particular scale of the culturing system. Method of co-culture systems including fungi and bacteria with filamentous microalgae [32, 75], and transgenic organisms [67] are challenging. Two-step approaches for optimization of growth and lipid accumulation are applied and showed robust results [12]. The authors Acién et al. showed analysis of costs where utilization of WW could reduce the total production costs to \$ 1.5 kg^{-1}, and using flue gases from industrial sources can decrease the cost of microalgae production to \$ 2.08 kg^{-1} [2]. There are examples of successful application of WW as follows: piggery, municipal, and monosodium glutamate ones under mixotrophic cultivation of microalgae [4, 49, 212, 213]. Effect of organic carbon to nitrogen ratio in WW on growth, nutrient uptake, and lipid accumulation of mixotrophic microalgae *Chlorella* sp. [164, 199] in microalgae culturing using non-sterile domestic WW. The application of advanced new methods (ultrasonic) also takes place in the state of the art in the area [165]. Therefore, already it has been proved that the use of WW to produce biofuels from microalgae showed cost-effectiveness and is one alternative solution for removal of a wide range of pollutants and recycling of water which is crucial for environmental safety. It must be noticed that optimal overall process of BD production can be achieved when it includes all stages and unit operations including PBRs design and harvesting of biomass from culture medium. Further, the cost of extraction and purification of compounds of interests must be minimized for the given scale of technology for BD production.

7 Key State Parameters of the Microalgae Process—Light

Light is a key state parameter of the microalgae cultivation and photosynthesis process. In some species, changes of photon flux density (PFD) resulted in suppression of CO_2 concentrating mechanism (CCM) by demining CO_2 transport capacity [190]. Optimization of light absorption for best cultivation development and modeling of the mechanisms involved are unavoidable [106, 123, 150, 234]. Valuable work and demonstration of how to improve cultivation conditions and PBR performance by using kinetic models recently was published by the authors [53]. In their studies, nine kinetics models of specific growth rate (SGR) were evaluated for both state parameters, intensity of light and its influence on biomass synthesis. The predictive power of the models was checked out on sets of independent experimental data, and Bannister and Grima models performed best compared to the other ones. As all kinetics models on population level in microalgae process limitations exist and must be taking into consideration such as: light spectra, flashing light effects (FLE), and light/dark cycles. Modification of the models reflecting in enhancement of their accuracy for different strain and conditions in PBRs for every particular case is necessary. Further, benefits are connected to application of the models for scale-up purposes and achieving of high-density biomass and high-value metabolites. Therefore, light conditions in different scenarios such as high light intensity (HL) resulted to high lipid and preferable triacylglycerol (TAG) overproduction [11, 45, 65, 169, 117, 113]. The light spectrum also affects synthesis of microalgae algae metabolites. For example, lipid accumulation [13] and many studies investigated the impact of light emitting diodes (LED), especially the influence of blue light [11, 232]. Some species are sensitive to the green light intensity which reflected the lipid content [74]. Studies showed that light with different wavelengths and light sources affected the cell growth, fatty acids content, and other microalgae metabolites [86, 108, 122]. In conclusion, the above analysis of light as a key criterion in microalgae process development demonstrated that light intensity, light spectrum, and pulsing light crucially controlled cell metabolism and its products. Light solely cannot be used as a control parameter for achieving a competitive microalgae process with high lipid content and biomass production. Temperature and nutrients must be evaluated alone empirically and further studied in combination with other environmental factors by using mathematical modeling.

8 Temperature

In focusing on the effect of temperature on lipid accumulation in different microalgae species, the assessment made cannot be unambiguous about the cellular response. For example, in *Tetraselmis subcordiformis*, *N. oculata*, the cells response under different temperature influence on lipid accumulation is not the same [216]. In another study, it was observed that decreasing the temperature of the process, the levels of different

FAs were changed [33, 236]. It has to be pointed out that for some species like *Scenedesmus sp.* [225], *Chlorella minutissima* [19], and *Chlamydomonas reinhardtii* [76], the cell growth and lipid production as a function of temperature exponentially increase to a certain extent. Therefore, the lipid content and FA composition of microalgae can be influenced to a certain extent by temperature level, but using as a single control parameter cannot be effective.

9 PH

The pH is a key state parameter of microalgae cultivation because its value controls solubility of minerals and dissociation of CO_2 in the medium [157]. Factors such as buffers, load of CO_2, temperature, and especially growth of the cells (producing OH^- ions) change the pH value [181]. The pH optimal value in most freshwater microalgae is in the range from 6 to 8 [159, 235]. Interestingly, the microalgae tolerate growth optimality in the wide pH intervals, but beyond it the yield of biomass decreases. Different studies showed different optimal ranges of pH values (7.0–10.0). It must be cited that some species have optimal pH below 3.0 [119, 221]. For *Scenedesmus almeriensis*, the optimal pH range is from 7.5 to 8.5. An expected fact is that the biomass productivity drops dramatically at pH above 9.0 [173]. During the photosynthetic process of CO_2 sequestration, the hydroxide ion (OH^- continuously is liberated by the cells and accumulates in cultural medium; hence the value of pH will increase if no buffers are used [167]. It means the chemical equilibrium of the inorganic carbon dissociated species of dissolved CO_2 shifts to carbonate ion (CO_3^-). It must be noticed the growth of microalgae on this ion is minimum [118]. On the other hand, lowering pH of the medium directs the chemical equilibrium to the dissolved CO_2 form. This phenomenon depends on the dynamics of gas loading, growth, and low biomass concentration and may lead to the release of big percentages of CO_2 into the gas phase which can jeopardize the effectiveness of the cultivation process. In processes of anaerobic digestion of pollutants, if nitrogen is supplied as ammonium, the pH of the solution increases depending on the load [18, 66, 118]. Studying thermodynamic properties of ammonium as a function of pH changes shows that in high values of pH, ammonia is released into the atmosphere depending on mixing condition and decrease in the liquid phase which resulted in nitrogen depletion and the growth of microalgae cells decreased. The same phenomenon may occur with phosphorus concentration in elevated pH, because there is a possibility of precipitation of phosphate in the forms as calcium phosphate, iron phosphate, and aluminum phosphate. Therefore, this can be a limiting condition of growth [18, 214]. At the molecular level, the changes of pH may affect the pH of microalgae cytoplasm which is neutral or slightly alkaline. At the population level, there are lots of studies about influence of pH and other key factors on the biomass growth. For example, cultivation of *Scenedesmus almeriensis* in continuous cultures is one of them [24]. Therefore, pH monitoring and optimal control will prevent the collapse of culture by avoiding critical pH values [77]. Tripathi et al. [206] studied the effect of pH on the growth of *Scenedesmus sp.* in a pH range from 7 to 10 and concluded that the optimal pH for this species was equal to 8. Munir et al. [141] investigated the growth behavior of the microalgae species (*Spirogyra sp.* and *Oedogonium sp.*) in pH values between 6.5 and 9.0. Interestingly, maximum growth was reached at pH 7.5 for both species. Wu et al. chose as object *Scenedesmus sp. LX1* and performed the experiments in the pH range from 5 to 11 [223]. The results showed that no difference of growth was observed for pH values of 7, 9, and 11. Lowering the pH to 5 decreased the cells growth. Hence, dynamic changes of pH values during the cultivation must be avoided. It can be done by adding proper buffers. Controlling the pH of the medium easily can be done by loading CO_2-enriched air [210]. Such a strategy was demonstrated by authors [157]. They showed in detail how *Chlorella sorokiniana DOE1412* strain responds during the CO_2 addition in PBR cultivation at different pH values (pH opt = 8.5). However, their findings showed that the lipid content was not a function of pH in the PBRs and, overall, the lipid content was slightly lower at the larger scale—25.7%. These data were similar to the results of Moheimani N. R. [135].

Dissolved CO_2 acidifies the liquid phase but during its consumption by the cells, the medium pH increases because of the OH^- ion liberated by the cells [25]. Moreover, special attention has to be given to the nitrogen source when pH is a control parameter of the process. In any case, pH will be affected by the cell growth as a function of uptake of the CO_2 as inorganic carbon substrate. Strategy of pH regulation has to focus on the optimal value favoring maximum biomass productivity and lipid synthesis.

10 Nutrients as Key Factors for Maximization of Lipid Synthesis

Firstly, microalgae cells have different needs for nutrient medium components. Mineral requirements can be calculated in both ways: direct measurements of their utilization by the cell elemental composition analysis [94]. This is fundamental for all macroelements [94, 156] to determine the perfect medium composition (without nutrients limitations) which is a requirement for realization of cost-effective industrial processes. Micronutrients can be determined by using proper analytical methods [34]. Extremely useful is to express chemical components of the cells by stoichiometric equations. It was shown excellently in the early original work of J. A. Roels, and during 40 years his knowledge was applied in the development of a wide range of biotechnological processes [168]. Therefore, application of stoichiometry for microalgae cells is beneficial. For example, for the two freshwater species *Chlamydomonas reinhardtii* (Eq. 18) and *Neochloris oleoabundans* (Eq. 19), stoichiometric equations can be written as follows:

$$CO_2 + 0.593H_2O + 0.176\,NH_4^+ + 0.007\,SO_4^{2-} + 0.018\,PO_4^{3-} - >$$

$$CH_{1.781}O_{0.437}N_{0.176}S_{0.007}P_{0.018} + 0.108\ H^+ + 1.127O_2 \tag{18}$$

$$CO2 + 0.751\ H_2O + 0.148\ NO_3^- + 0.014\ SO_4^{2-} + 0.012\ PO_4^{3-} + 0.212H^+ - >$$

$$CH_{1.715}O_{0.427}N_{0.148}S_{0.014}P_{0.012} + 1.437\ O_2 \tag{19}$$

On the basis of them, researchers can calculate yields, microalgae growth, and substrate and product rates as well as cell needs of nitrogen-, phosphorus-, and sulfur sources during photosynthesis. Analysis of Eqs. 1 and 2 shows how much CO_2 microalgae needs and respective macroelements [47].

Always, it is valuable to determine minimum nutritional requirements needed for the growth of microalgae. It can be done by using the approximate molecular formula $CO_{0.48}H_{1.83}N_{0.11}P_{0.01}$ [1] where the one carbon atom is the referent point. This stoichiometric formula shows that the most important macronutrients for autotrophic growth are the carbon (C), nitrogen (N), and phosphorus (P) [167]. Hence, in the molecular formula of biomass the C represents 50% [1]. Supply of carbon is needed in high concentrations because it is vital for synthesis of cell components—carbohydrates, proteins, nucleic acids, vitamins, lipids, etc. [167]. Transport of inorganic carbon in algae can be done by diffusion ($5.0 < pH < 7.0$) and active transport ($pH > 7.0$) [63]. CO_2 and bicarbonates (HCO_3^-) supply are very important in maximization of autotrophic growth and product formation [1]. Mixotrophic [7, 17, 155, 172] and heterotrophic cultivation [23, 105, 149] by using sugars, acids, alcohols, etc. especially beneficial for microalgae heterotrophic and mixotrophic growth is the use of WW serving as a good source of organic carbon. Here, we will focus on nutrients which can be used as a stress factor for induction of higher lipid synthesis in microalgae**.** Nitrogen is one of them.

11 Nitrogen

Nutrients can be used as a stress factor to promote synthesis of secondary metabolites. In this case nitrogen depletion induces lipid synthesis in microalgae. From experimental data and stoichiometry, nitrogen value in cells can be from 1 up to 10% (dry matter weight) [220] and is obtained from chemical compounds—ammonia, nitrates, and nitrites. In metabolic pathways, nitrogen is involved in amino acid synthesis and of TAGs, proteins, and nucleic acids production. It also helps the work of light harvesting complexes in the chloroplasts. Nitrogen stress reduces microalgae growth and photosynthetic activity, but induces lipid accumulation [57, 61, 112, 137].

It is proven that nitrogen starvation serves as a booster of lipid augmentation in many algae species [142]. Nitrogen deficiency serves as a significant stimulus for lipid accumulation of industrially important microalgae from the species *Chlorella zofingiensis* [116] and *Scenedesmus quadricauda* [6]. These reports imply that the degradation of nitrogen compounds is most probably the reason for lipid accumulation under nitrogen deficiency conditions. On the other hand, researchers must be careful with nitrogen supplementation by aiming to increase the biomass growth of algae because there is a study showing that over-dosage of nitrogen leads to ammonia toxicity and cell death [38].

12 Phosphate

Another stress factor of the nutrient medium can be phosphorus deficiency. When applied as a technique of control alone or in combination [31, 111] for the cultivation process, it can increase to some extent the accumulation of lipids and FAs [126, 127, 133, 177, 186, 227]. It is interesting finding that the changes of FA composition were similar to the ones achieved under nitrogen deficiency. It is obvious that phosphorus deficiency stress alone can impact cell lipid synthesis much smaller compared to nitrogen deficiency conditions because of the different roles which these elements play in algae metabolism [56, 126, 127].

Trace elements in the medium can also be used as stress factors to induce an elevated lipid synthesis [162, 198].

13 Salinity

There are lots of studies showing the effectiveness of saline stress as a methodology in culturing microalgae to induce synthesis of lipids which are considered in biodiesel/biofuel technologies. For example, the lipid accumulation in *C. reinhardtii, A. dimorphus, C. vulgaris,* and *Acutodesmus obliquus* is observed when a concentration of NaCl above 0.1 M is added [3, 71, 146]. Different cultivation techniques can be applied as well, to induce salinity stress [51, 54, 109, 183]. During cultivation of microalgae species on WW, the combination with salt stress is a plus [143, 144]. Microalgal metabolism of lipids is influenced by other salts such as KCl, $MgCl_2$, and $CaCl_2$ and may compete for better effectiveness than NaCl [192]. It is predictable that saline stress will change to some extend the composition of FAs [128, 228]. Further, a high concentration of NaCl can elevate MUFA profile and increase the SFA in these conditions [10, 70, 146]. Of course, the mechanisms of synthesis of FAs under environmental stress and growth can be analyzed systematically [54, 196]. In conclusion, the role of saline stress in biodiesel process development from microalgae has to be studied in detail and optimally designed in every particular case when using fresh- and salt water microalgae.

14 Carbon Sources Affecting Lipid Accumulation—CO_2

The use of CO_2 from the atmosphere and industrial waste gases must be evaluated through the water chemistry of this gas and its thermodynamic properties. The enzyme carbonic anhydrase, as the main transporter responsible for the cellular absorption of CO_2, has always been a scientific target for improving its efficiency to catch a dissociated CO_2 in a form of bicarbonate ion (HCO_3^-), which is taken up by Rubisco to fulfill the cell metabolism. Research in the area showed that a highly active carbonic anhydrase gene taken from *Mesorhizobium loti* was coded in *Chlorella sorokiniana* and *Chlorella vulgaris*. The results were promising because enzyme activity was of 31% and 23% higher compared to the wild strains [115]. Interesting findings about carbonic anhydrase manipulation were when it was taken from *Sulfurihydrogenibium yellowstonense* and transformed into *Chlamydomonas reinhardtii CC-400*. The new modified strain utilized CO_2 of 2.8 g-CO_2/g_{DW} under mixotrophic condition and 5% CO_2 supply, which is a remarkable achievement [114]. Mixotrophic growth of microalgae demonstrated great potential for utilization of CO_2 in this cultivation mode. Hence, the introduction of glucose transporter genes widens the horizons of carbon capture mechanisms. Kramer et al. expressed heterologous hexose uptake protein (HUP1) from *Chlorella kessleri* into *Phaeodactylum tricornutum*, in order to enhance glucose utilization and consequently biomass concentration under low light intensities [88]. Research optimizing genetic manipulation techniques and utilizing enzymes from various microorganisms keeps alive the idea to explore the full potential of microalgae for maximum CO_2 fixation and high value product formation.

15 $NaHCO_3$ Concentration

$NaHCO_3$ is extremely important source of inorganic carbon for autotrophic microalgae and from a techno-economical consideration is more efficient than gaseous CO_2 in microalgae cultivation. Its transportation and storage are costly in comparison with bicarbonate. It has to be pointed out that the cultivation of microalgae with bicarbonate as a buffer [69, 231] as well as a carbon source has the distinct advantages of high carbon utilization efficiency because its application can serve two goals: buffering the cultural medium and being a depot of CO_2 capture from gas phase which is continuously loaded in the PBR vessel.

In recent studies, application of $NaHCO_3$ has been found to significantly increase the total lipid content of *Scenedesmus, Nannochloris, D. salina,* and *D. tertiolecta.* The addition of bicarbonate in combination with nutrient deficiency increased additionally the lipid content of *D. salina V-101* [182, 191]. Remarkably, in some microalgae species, $NaHCO_3$ is the preferred carbon source, in combination with other factors, than CO_2 especially for lipid biosynthesis [147]. High concentrations of bicarbonate increased the activity of Rubisco enzyme by helping the synthesis of 3-phosphoglycerate, which is a precursor for FA production [148].

Further achievements in the state of the art by exploring the bicarbonate features and effect on lipid synthesis is seen in the original work of authors [84]. They demonstrated a new method of culturing algae cells where water electrolysis products are utilized, and the final biomass concentration reached approximately three-fold higher compared with the simple bicarbonate addition. The authors claimed that such an approach has several advantages connected with operational conditions which are guided by simple electrical signals avoiding sophisticated gas facilities. Moreover, knowledge about water electrolysis is big enough to enable easier implementation of this system in process development and solving scale-up problems. Of course, this promising method requires future work for optimization of electrolytic cell design and state parameters in any particular case and algae strain choice.

16 Wastewater as a Source of Nutrients for Biodiesel Production in Microalgae Technology

In many places of this chapter, we pointed out the importance of WW utilization for cost-effective BD production from microalgae. An excellent systematic review on this subject can be found elsewhere [179]. The authors analyzed application of microalgae technology in an integrated system where WW must be purified and moreover to be used as valuable resources of nutrients for microalgae. Hence, nutrient-rich WW should be utilized via biological uptake through microalgae biomass and its consortium with other microorganisms. Ability of microalgae cells to utilize organic carbon, nitrogen, and phosphorus is very high by producing biomass containing high value products for bioenergy and many other applications. The authors used a systematic approach to provide guidance toward implementing an integral concept for WW system nutrient and turbidity control, discussing available methods for pre-treatments which helped to make WW suitable media for microalgae cultivation.

17 PBRs Design and Functionality as a System

During the last several decades, closed PBRs were investigated extensively. All types, column, tubular, airlift, flat plate, pyramid, and hybrid (with membrane, etc.), are discussed in detail in the literature. Recently, an original review on the subject was published by considering advantages, disadvantages, challenges, and future prospects of closed PBR systems [20]. We are going to share our 40-years' experience in bioreactor modeling and especially on the subject application of our knowledge which for the first time demonstrated a systematic approach for mathematical analysis, design, and scale-up of column PBRs [95]. We demonstrated the power of this approach when analyzing the pandemic in 2020 [97, 175].

Here, we are going to briefly analyze column PBR as a system by using mathematical modeling in the frame of system analysis theory [80–82] and how this approach helps researchers to understand the complexity of it and internal relationships among parallel physical, chemical, biochemical, and microbiological processes taking place in the closed PBRs. It can be done when PBR as a system is divided to sub-systems which are studied separately. The sub-systems are arranged in hierarchic levels. Further, interrelationships between levels are studied and modeled. All the models of sub-systems are combined in a simple form in order to be useful when building the complex mathematical model of PBR. The most important sub-system is the kinetics of photosynthesis, and it stays as a base in the frame of hierarchy of sub-systems and determines the directions of optimal design and process development in column PBR. The other key sub-system considered is the hydrodynamics which linked with other sub-systems and is fundamentally important. The tool as the computational fluid dynamics CFD [69] builds the base of complex and accurate modeling with strong connection with microalgae growth and kinetics. It must be noticed that our successful experience in laboratories from Russia, Bulgaria, Brazil, and the USA highlighted the challenging horizons of the complex modeling of biotechnological processes [218, 219] as well as the WW treatment processes [15, 42, 59, 90, 130, 134, 136, 203, 204], purification of heavy metals [15], bioreactors [90, 92], PBRs [95], microalgae growth kinetics [35, 36, 91, 231], and bacterial growth kinetics [202]. Important is to mention that bioenergy concepts for optimal ethanol production processes were successfully modeled as well [89, 93]. The fundamentals of microalgae and biotechnological laboratories are in the frame of integral biorefinery concept for complete utilization of the algae biomass. As described above, special attention must be paid to strain selection with the potential of synthesis of high value products, for example lipids.

Therefore, the strategy of system analysis theory is a robust and winning one for process development, design, and scale-up of any construction of closed PBRs. A robust novel complex PBR phenomenological model and a light model linked with algal physiology can be successfully developed by using this theory as a basis [95]. As is mentioned above, the system analysis theory was applied to evaluate the CO_2 sequestration from flue gas by using microalgae. Moreover, the algorithm for development of a complex PBR phenomenological model can be a tool for understanding and sharing of knowledge between members of a multidisciplinary team, and as a result it can be accepted as a common strategy for microalgae overall process development and maximization of biomass and high value products from any chosen strain of interest. In particular, the application of this strategy to a maximization of lipids for biodiesel production will be extremely valuable.

18 Limitations and Future Perspectives of Closed PBRs

The innovative approach for creation of PBRs must consider as well cost-effectiveness of this close system. Techno-economical analysis of cost requires from scientists to look for solutions to minimize the costs of their construction, maintenance, and overall use in the utilization of CO_2 from industrial gases.

This example proved the idea. The cost of 1 kg of algae production from raceways ponds is estimated to be about $16.4; in contrast the cost algae production in PBR is to be estimated around $33 [16]. About the usage of land, it is estimated that production of 1 kg of algae biomass needs about 1 acre, and this will cost approximately $11. It must be noticed that the scale is a key and crucial parameter. An application of high advanced technologies (for 100 acres) is estimated to minimize the cost of production of about $4.4 [41].

Therefore, with the limitation over closed PBR design, the techno-economical analysis for implementation of PBRs in microalgae technology must be evaluated through the dynamics of cost changes as a function of scale-up procedure. It is valid for any microalgae technology.

19 Downstream Processes for Lipid Isolation from Microalgae Biomass

Once again, from a techno-economical point of view, the cost of downstream processes of any targeted metabolites plays a crucial role in the determination of the economics and of the process realization. The steps such as dewatering, concentration, cell disruption, purification, and extraction of the desirable metabolite as a product remain serious bottlenecks to overcome in the microalgae cultivation technology [85, 103, 132]. Specific characteristics of microalgae metabolites determined the variety of methods for their identification and isolation. In current research, advanced techniques for biometabolite extraction, separation, and purification from algae for biofuel production in the frame of an integral biorefinery approach can be found elsewhere [50, 131, 138, 189, 197, 200].

It must be emphasized that genetic engineering of microalgae to increase lipid production, applications of transformation techniques, and genome editing tools will dramatically change the horizons of microalgae potential to produce desired metabolites, in this case, lipids and FAs as precursors for biodiesel production. Of course, using these modern tools must be done with cautions as well as selection of target mutants because of the characteristics and some specific features of genetically engineered strains of microalgae.

20 Future Outlook

Current data highlight the obtained achievement isolation of highly effective microalgae strains producing high value products. As always, definitive decisions are made from economical visibility of the final product [161]. Optimization of upstream (where special attention is given on optimal PBR design) and downstream steps [28, 29] needs the application of modern methods for search of maximum productivity or minimizing the cost by applying dynamic programming where the scheme of microalgae technology can achieve cost-effectiveness in the market. In addition, genetically modified microalgae strains will be a strong support to change the microalgae potential as a bioenergy source.

21 Conclusions

Microalgae lipids are excellent alternative as biodiesel precursors, but fuel obtained from microalgae must overcome financial restriction in order to be cost-effective on the market. Microalgae growth has to be optimal in the chosen construction of closed innovative PBRs which demonstrate the importance of selection of microalgae strains accumulating bioenergetics products in the biomass. If possible, all key state parameters of the process must be optimal and nutrients in the medium should be in access supporting non-limiting conditions of growth and product formation. Of course, the process optimality is a compromise between maximum productivity and cost-effectiveness of overall production scheme. Using different sources of nutrients from wastewaters will help to meet environmentally clean technology requirements.

Acknowledgements This work was supported by Grant КП-06-Н53/2 (11.11.2021) from Bulgarian National Science Fund which is gratefully acknowledge.

References

1. Acién Fernández FG, Fernández Sevilla JM, Molina Grima E (2013) Photobioreactors for the production of microalgae. Rev Environ Sci Bio/Technol 12(2):131–151. https://doi.org/10.1007/s11157-012-9307-6
2. Acién FG, Fernández JM, Magán JJ, Molina E (2012) Production cost of a real microalgae production plant and strategies to reduce it. Biotechnol Adv 30(6):1344–1353. https://doi.org/10.1016/j.biotechadv.2012.02.005
3. Ahmed RA, He M, Aftab RA, Zheng S, Nagi M, Bakri R, Wang C (2017) Bioenergy application of dunaliella salina SA 134 grown at various salinity levels for lipid production. Sci Rep 7(1):8118. https://doi.org/10.1038/s41598-017-07540-x
4. Aketo T, Hoshikawa Y, Nojima D, Yabu Y, Maeda Y, Yoshino T, Takano H, Tanaka T (2020) Selection and characterization of microalgae with potential for nutrient removal from municipal wastewater and simultaneous lipid production. J Biosci Bioeng 129(5):565–572. https://doi.org/10.1016/j.jbiosc.2019.12.004

5. Al-lwayzy SH, Yusaf T, Al-Juboori RA (2014) Biofuels from the fresh water microalgae chlorella vulgaris (FWM-CV) for diesel engines. Energies 7:1829–1851. https://doi.org/10.3390/en7031829
6. Anand J, Arumugam M (2015) Enhanced lipid accumulation and biomass yield of scenedesmus quadricauda under nitrogen starved condition. Biores Technol 188:190–194. https://doi.org/10.1016/j.biortech.2014.12.097
7. Angelini F, Bellini E, Marchetti A, Salvatori G, Villano M, Pontiggia D, Ferrari S (2024) Efficient utilization of monosaccharides from agri-food byproducts supports chlorella vulgaris biomass production under mixotrophic conditions. Algal Res 77:103358. https://doi.org/10.1016/j.algal.2023.103358
8. Arora N, Patel A, Mehtani J, Pruthi PA, Pruthi V, Poluri KM (2019) Co-culturing of oleaginous microalgae and yeast: paradigm shift towards enhanced lipid productivity. Environ Sci Pollut Res 26(17):16952–16973. https://doi.org/10.1007/s11356-019-05138-6
9. Arora N, Philippidis GP (2021) Insights into the physiology of chlorella vulgaris cultivated in sweet sorghum bagasse hydrolysate for sustainable algal biomass and lipid production. Sci Rep 11(1):6779. https://doi.org/10.1038/s41598-021-86372-2
10. Atikij T, Syaputri Y, Iwahashi H, Praneenararat T, Sirisattha S, Kageyama H, Waditee-Sirisattha R (2019) Enhanced lipid production and molecular dynamics under salinity stress in green microalga chlamydomonas reinhardtii (137C). Marine Drugs 17. https://doi.org/10.3390/md17080484
11. Atta M, Idris A, Bukhari A, Wahidin S (2013) Intensity of blue LED light: a potential stimulus for biomass and lipid content in fresh water microalgae chlorella vulgaris. Biores Technol 148:373–378. https://doi.org/10.1016/j.biortech.2013.08.162
12. Aziz MMA, Kassim KA, Shokravi Z, Jakarni FM, Liu HY, Zaini N, Tan LS, Islam ABMS, Shokravi H (2020) Two-stage cultivation strategy for simultaneous increases in growth rate and lipid content of microalgae: a review. Renew Sustain Energy Rev 119:109621. https://doi.org/10.1016/j.rser.2019.109621
13. Bialevich V, Zachleder V, Bišová K (2022) The effect of variable light source and light intensity on the growth of three algal species. Cells 11. https://doi.org/10.3390/cells11081293
14. Binnal P, Babu PN (2017) Statistical optimization of parameters affecting lipid productivity of microalga chlorella protothecoides cultivated in photobioreactor under nitrogen starvation. S Afr J Chem Eng 23:26–37. https://doi.org/10.1016/j.sajce.2017.01.001
15. Borba CE, Silva EAD, Fagundes-Klen MR, Kroumov AD, Guirardello R (2008) Prediction of the copper (II) ions dynamic removal from a medium by using mathematical models with analytical solution. J Hazard Mater 152(1):366–372. https://doi.org/10.1016/j.jhazmat.2007.07.005
16. Bošnjaković M, Sinaga N (2020) The perspective of large-scale production of algae biodiesel. Applied Sci 10. https://doi.org/10.3390/app10228181
17. Braun JCA, Balbinot L, Beuter MA, Rempel A, Colla LM (2024) Mixotrophic cultivation of microalgae using agro-industrial waste: tolerance level, scale up, perspectives and future use of biomass. Algal Res 80:103554. https://doi.org/10.1016/j.algal.2024.103554
18. Cai T, Park SY, Li Y (2013) Nutrient recovery from wastewater streams by microalgae: status and prospects. Renew Sustain Energy Rev 19:360–369. https://doi.org/10.1016/j.rser.2012.11.030
19. Cao J, Yuan H, Li B, Yang J (2014) Significance evaluation of the effects of environmental factors on the lipid accumulation of chlorella minutissima UTEX 2341 under low-nutrition heterotrophic condition. Biores Technol 152:177–184. https://doi.org/10.1016/j.biortech.2013.10.084
20. Chanquia SN, Vernet G, Kara S (2022) Photobioreactors for cultivation and synthesis: specifications, challenges, and perspectives. Eng Life Sci 22(12):712–724. https://doi.org/10.1002/elsc.202100070
21. Chen CY, Lee MH, Dong CD, Leong YK, Chang JS (2020a) Enhanced production of microalgal lipids using a heterotrophic marine microalga Thraustochytrium sp. BM2. Biochem Eng J 154:107429. https://doi.org/10.1016/j.bej.2019.107429

22. Chen CY, Lee MH, Leong YK, Chang JS, Lee DJ (2020b) Biodiesel production from heterotrophic oleaginous microalga Thraustochytrium sp. BM2 with enhanced lipid accumulation using crude glycerol as alternative carbon source. Bioresource Techn 306:123113. https://doi.org/10.1016/j.biortech.2020.123113.
23. Chen J-H, Huang J-X, Nagarajan D, Chang J-S (2024) Reutilization of food industrial waste for lutein production with heterotrophic microalgae chlorella sorokiniana MB-1-M12. J Taiwan Inst Chem Eng 160:105243. https://doi.org/10.1016/j.jtice.2023.105243
24. Chiranjeevi P, Mohan SV (2016) Critical parametric influence on microalgae cultivation towards maximizing biomass growth with simultaneous lipid productivity. Renewable Energy 98:64–71. https://doi.org/10.1016/j.renene.2016.03.063
25. Chiu S-Y, Kao C-Y, Chen C-H, Kuan T-C, Ong S-C, Lin C-S (2008) Reduction of CO_2 by a high-density culture of chlorella sp. in a semicontinuous photobioreactor. Biores Technol 99(9):3389–3396. https://doi.org/10.1016/j.biortech.2007.08.013
26. Cho D-H, Ramanan R, Heo J, Lee J, Kim B-H, Oh H-M, Kim H-S (2015) Enhancing microalgal biomass productivity by engineering a microalgal–bacterial community. Biores Technol 175:578–585. https://doi.org/10.1016/j.biortech.2014.10.159
27. Cho K, Hur S-P, Lee C-H, Ko K, Lee Y-J, Kim K-N, Kim M-S, Chung Y-H, Kim D, Oda T (2016) Bioflocculation of the oceanic microalga dunaliella salina by the bloom-forming dinoflagellate Heterocapsa circularisquama, and its effect on biodiesel properties of the biomass. Biores Technol 202:257–261. https://doi.org/10.1016/j.biortech.2015.12.047
28. Chong JWR, Khoo KS, Chew KW, Ting H-Y, Show PL (2023) Trends in digital image processing of isolated microalgae by incorporating classification algorithm. Biotechnol Adv 63:108095. https://doi.org/10.1016/j.biotechadv.2023.108095
29. Chong JWR, Tang DYY, Leong HY, Khoo KS, Show PL, Chew KW (2023) Bridging artificial intelligence and fucoxanthin for the recovery and quantification from microalgae. Bioengineered 14(1):2244232. https://doi.org/10.1080/21655979.2023.2244232
30. Chong JWR, Yew GY, Khoo KS, Ho S-H, Show PL (2021) Recent advances on food waste pretreatment technology via microalgae for source of polyhydroxyalkanoates. J Environ Manage 293:112782. https://doi.org/10.1016/j.jenvman.2021.112782
31. Chu F, Cheng J, Li K, Wang Y, Li X, Yang W (2020) Enhanced lipid accumulation through a regulated metabolic pathway of phosphorus luxury uptake in the microalga chlorella vulgaris under nitrogen starvation and phosphorus repletion. ACS Sustain Chem and Eng 8(22):8137–8147. https://doi.org/10.1021/acssuschemeng.9b07447
32. Chu R, Li S, Zhu L, Yin Z, Hu D, Liu C, Mo F (2021) A review on co-cultivation of microalgae with filamentous fungi: efficient harvesting, wastewater treatment and biofuel production. Renew Sustain Energy Rev 139:110689. https://doi.org/10.1016/j.rser.2020.110689
33. Chua ET, Dal'Molin C, Thomas-Hall S, Netzel ME, Netzel G, Schenk PM (2020) Cold and dark treatments induce omega-3 fatty acid and carotenoid production in Nannochloropsis oceanica. Algal Res 51:102059. https://doi.org/10.1016/j.algal.2020.102059
34. Cogne G, Lehmann B, Dussap C-G, Gros J-B (2003) Uptake of macrominerals and trace elements by the cyanobacterium Spirulina platensis (Arthrospira platensis PCC 8005) under photoautotrophic conditions: culture medium optimization. Biotechnol Bioeng 81(5):588–593. https://doi.org/10.1002/bit.10504
35. Crofcheck C, Montross M, Cassidy K, Kroumov A et al (2010) Medium optimization for CO_2 mitigation from flue gas. Annual Institute of Biological Engineering Meeting in Cambridge, Cambridge, MA
36. Crofcheck CL, Monstross M, Xinyi E, Shea AP, Crocker M, Andrews R (2012) Influence of media composition on the growth rate of chlorella vulgaris and Scenedesmus acutus utilized for CO_2 mitigation. In: 2012 Dallas, Texas, July 29–August 1, 2012. St. Joseph, MI, ASABE. https://doi.org/10.13031/2013.41734
37. D'Alessandro EB, Antoniosi Filho NR (2016) Concepts and studies on lipid and pigments of microalgae: a review. Renew Sustain Energy Rev 58:832–841. https://doi.org/10.1016/j.rser.2015.12.162

38. Dai G-Z, Qiu B-S, Forchhammer K (2014) Ammonium tolerance in the cyanobacterium ynechocystis sp. strain PCC 6803 and the role of the multigene family. Plant, Cell Environ 37(4):840–851. https://doi.org/10.1111/pce.12202
39. Damiani MC, Popovich CA, Constenla D, Leonardi PI (2010) Lipid analysis in Haematococcus pluvialis to assess its potential use as a biodiesel feedstock. Biores Technol 101(11):3801–3807. https://doi.org/10.1016/j.biortech.2009.12.136
40. Darnoko D, Cheryan M (2000) Kinetics of palm oil transesterification in a batch reactor. J Am Oil Chem Soc 77(12):1263–1267. https://doi.org/10.1007/s11746-000-0198-y
41. Davis R, Aden A, Pienkos PT (2011) Techno-economic analysis of autotrophic microalgae for fuel production. Appl Energy 88(10):3524–3531. https://doi.org/10.1016/j.apenergy.2011.04.018
42. de Pauli AR, Espinoza-Quiñones FR, Trigueros DEG, Módenes AN, de Souza ARC, Borba FH, Kroumov AD (2018) Integrated two-phase purification procedure for abatement of pollutants from sanitary landfill leachates. Chem Eng J 334:19–29. https://doi.org/10.1016/j.cej.2017.10.028
43. Deng X-Y, Gao K, Addy M, Li D, Zhang R-C, Lu Q, Ma Y-W, Cheng Y-L, Chen P, Liu Y-H, Ruan R (2018) Cultivation of chlorella vulgaris on anaerobically digested swine manure with daily recycling of the post-harvest culture broth. Biores Technol 247:716–723. https://doi.org/10.1016/j.biortech.2017.09.171
44. Diasakou M, Louloudi A, Papayannakos NG (1998) Kinetics of the non-catalytic transesterification of soybean oil. Fuel 77:1297–1302
45. Difusa A, Talukdar J, Kalita MC, Mohanty K, Goud VV (2015) Effect of light intensity and pH condition on the growth, biomass and lipid content of microalgae scenedesmus species. Biofuels 6(1–2):37–44. https://doi.org/10.1080/17597269.2015.1045274
46. Ding W, Zhao Y, Xu J-W, Zhao P, Li T, Ma H, Reiter RJ, Yu X (2018) Melatonin: a multifunctional molecule that triggers defense responses against high light and nitrogen starvation stress in haematococcus pluvialis. J Agric Food Chem 66(29):7701–7711. https://doi.org/10.1021/acs.jafc.8b02178
47. Doucha J, Straka F, Lívanský K (2005) Utilization of flue gas for cultivation of microalgae chlorella sp.) in an outdoor open thin-layer photobioreactor. J Appl Phycol 17(5):403–412. https://doi.org/10.1007/s10811-005-8701-7
48. Dutta K, Daverey A, Lin (2014) Evolution retrospective for alternative fuels: first to fourth generation. Renewable Energy 69(C):114–122. https://doi.org/10.1016/j.renene.2014.02
49. Ebrahimian A, Kariminia H-R, Vosoughi M (2014) Lipid production in mixotrophic cultivation of chlorella vulgaris in a mixture of primary and secondary municipal wastewater. Renewable Energy 71:502–508. https://doi.org/10.1016/j.renene.2014.05.031
50. Egesa D, Plucinski P (2024) Efficient extraction of lipids from magnetically separated microalgae using ionic liquids and their transesterification to biodiesel. Biomass Conversion and Biorefinery 14(1):419–434
51. El-Sheekh MM, Galal HR, Mousa ASHH, Farghl AAM (2024) Impact of macronutrients and salinity stress on biomass and biochemical constituents in Monoraphidium braunii to enhance biodiesel production. Sci Rep 14(1):2725. https://doi.org/10.1038/s41598-024-53216-8
52. Esakkimuthu S, Krishnamurthy V, Wang S, Hu X, Abomohra AEF (2020) Application of p-coumaric acid for extraordinary lipid production in Tetradesmus obliquus: a sustainable approach towards enhanced biodiesel production. Renewable Ener 157:368–376. https://doi.org/10.1016/j.renene.2020.05.005
53. Estevam BR, Pinto LFR, Filho RM, Fregolente LV (2024) Effects of cultivation systems and nutrient limitation on the growth and metabolite biosynthesis of Botryococcus terribilis. Biomass Convers Biorefinery (15)
54. Farkas A, Pap B, Zsíros O, Patai R, Shetty P, Garab G, Bíró T, Ördög V, Maróti G (2023) Salinity stress provokes diverse physiological responses of eukaryotic unicellular microalgae. Algal Res 73:103155. https://doi.org/10.1016/j.algal.2023.103155

55. Fawzy MA, El-Otify AM, Adam MS, Moustafa SSA (2021) The impact of abiotic factors on the growth and lipid accumulation of some green microalgae for sustainable biodiesel production. Environ Sci Pollut Res 28(31):42547–42561. https://doi.org/10.1007/s11356-021-13781-1
56. Feng P, Deng Z, Fan L, Hu Z (2012) Lipid accumulation and growth characteristics of chlorella zofingiensis under different nitrate and phosphate concentrations. J Biosci Bioeng 114(4):405–410. https://doi.org/10.1016/j.jbiosc.2012.05.007
57. Feng P, Deng Z, Hu Z, Wang Z, Fan L (2014) Characterization of Chlorococcum pamirum as a potential biodiesel feedstock. Biores Technol 162:115–122. https://doi.org/10.1016/j.biortech.2014.03.076
58. Ferreira GF, Ríos Pinto LF, Maciel Filho R, Fregolente LV (2019) A review on lipid production from microalgae: association between cultivation using waste streams and fatty acid profiles. Renew Sustain Energy Rev 109:448–466. https://doi.org/10.1016/j.rser.2019.04.052
59. Fiorentin LD, Módenes AN, Espinoza-Quiñones FR, Trigueros DEG, Kroumov AD, Manenti DR, Borba CE (2015) Biosorption of the reactive blue 5G dye in a fixed bed column packed with orange bagasse: experimental and mathematical modelling. Sep Sci Technol 50(15):2267–2275. https://doi.org/10.1080/01496395.2015.1047453
60. Freedman B, Butterfield RO, Pryde EH (1986) Transesterification kinetics of soybean oil. J Am Oil Chemists' Soc 63:1375–1380. https://doi.org/10.1007/BF02679606
61. Fu L, Cui X, Li Y, Xu L, Zhang C, Xiong R, Zhou D, Crittenden JC (2017) Excessive phosphorus enhances chlorella regularis lipid production under nitrogen starvation stress during glucose heterotrophic cultivation. Chem Eng J 330:566–572. https://doi.org/10.1016/j.cej.2017.07.182
62. Gim GH, Kim JK, Kim HS, Kathiravan MN, Yang H, Jeong S-H, Kim SW (2014) Comparison of biomass production and total lipid content of freshwater green microalgae cultivated under various culture conditions. Bioprocess Biosyst Eng 37(2):99–106. https://doi.org/10.1007/s00449-013-0920-8
63. Gonçalves AL, Pires JCM, Simões M (2017) A review on the use of microalgal consortia for wastewater treatment. Algal Res 24:403–415. https://doi.org/10.1016/j.algal.2016.11.008
64. Gui J, Chen S, Luo G, Wu Z, Fan Y, Yao L, Xu H (2021) Nutrient deficiency and an algicidal bacterium improved the lipid profiles of a novel promising oleaginous dinoflagellate, prorocentrum donghaiense, for biodiesel production. Appl Environ Microbiol 87(19):e01159-e1121. https://doi.org/10.1128/AEM.01159-21
65. Guo X, Su G, Li Z, Chang J, Zeng X, Sun Y, Lu Y, Lin L (2015) Light intensity and N/P nutrient affect the accumulation of lipid and unsaturated fatty acids by chlorella sp. Biores Technol 191:385–390. https://doi.org/10.1016/j.biortech.2015.04.014
66. Guštin S, Marinšek-Logar R (2011) Effect of pH, temperature and air flow rate on the continuous ammonia stripping of the anaerobic digestion effluent. Process Saf Environ Prot 89(1):61–66. https://doi.org/10.1016/j.psep.2010.11.001
67. Han X, Song X, Li F, Lu Y (2020) Improving lipid productivity by engineering a control-knob gene in the oleaginous microalga Nannochloropsis oceanica. Metablc Eng Commun 11:e00142. https://doi.org/10.1016/j.mec.2020.e00142
68. Hegel P, Martín L, Popovich C, Damiani C, Pancaldi S, Pereda S, Leonardi P (2017) Biodiesel production from neochloris oleoabundans by supercritical technology. Chem Eng Process 121:232–239. https://doi.org/10.1016/j.cep.2017.08.018
69. Hinterholz CL, Trigueros DEG, Módenes AN, Borba CE, Scheufele FB, Schuelter AR, Kroumov AD (2019) Computational fluid dynamics applied for the improvement of a flat-plate photobioreactor towards high-density microalgae cultures. Biochem Eng J 151:107257. https://doi.org/10.1016/j.bej.2019.107257
70. Hounslow E, Evans CA, Pandhal J, Sydney T, Couto N, Pham TK, Gilmour DJ, Wright PC (2021) Quantitative proteomic comparison of salt stress in chlamydomonas reinhardtii and the snow alga chlamydomonas nivalis reveals mechanisms for salt-triggered fatty acid accumulation via reallocation of carbon resources. Biotechnol Biofuels 14(1):121. https://doi.org/10.1186/s13068-021-01970-6

71. Hounslow E, Kapoore RV, Vaidyanathan S, Gilmour DJ, Wright PC (2016) The search for a lipid trigger: the effect of salt stress on the lipid profile of the model microalgal species chlamydomonas reinhardtii for Biofuels Production. Current Biotechnology 5(4):305–313. https://doi.org/10.2174/2211550105666160322234434
72. Hu J, Wang D, Chen H, Wang Q (2023) Advances in genetic engineering in improving photosynthesis and microalgal productivity. Int J Mol Sci 24. https://doi.org/10.3390/ijms24031898
73. Hu Q, Sommerfeld M, Jarvis E, Ghirardi M, Posewitz M, Seibert M, Darzins A (2008) Microalgal triacylglycerols as feedstocks for biofuel production: perspectives and advances. Plant J 54(4):621–639. https://doi.org/10.1111/j.1365-313X.2008.03492.x
74. Hultberg M, Jönsson HL, Bergstrand K-J, Carlsson AS (2014) Impact of light quality on biomass production and fatty acid content in the microalga chlorella vulgaris. Biores Technol 159:465–467. https://doi.org/10.1016/j.biortech.2014.03.092
75. Huo S, Basheer S, Liu F, Elshobary M, Zhang C, Qian J, Xu L, Arslan M, Cui F, Zan X, Zhu F, Zou B, Ding Q, Ma H (2020) Bacterial intervention on the growth, nutrient removal and lipid production of filamentous oleaginous microalgae tribonema sp. Algal Res 52:102088. https://doi.org/10.1016/j.algal.2020.102088
76. James GO, Hocart CH, Hillier W, Price GD, Djordjevic MA (2013) Temperature modulation of fatty acid profiles for biofuel production in nitrogen deprived chlamydomonas reinhardtii. Biores Technol 127:441–447. https://doi.org/10.1016/j.biortech.2012.09.090
77. Jia H, Yuan Q (2016) Removal of nitrogen from wastewater using microalgae and microalgae–bacteria consortia. Cogent Environmental Science 2(1):1275089. https://doi.org/10.1080/23311843.2016.1275089
78. Jiang L, Pei H, Hu W, Han F, Zhang L, Hou Q (2015) Effect of diethyl aminoethyl hexanoate on the accumulation of high-value biocompounds produced by two novel isolated microalgae. Biores Technol 197:178–184. https://doi.org/10.1016/j.biortech.2015.08.068
79. Jiao H, Tsigkou K, Elsamahy T, Pispas K, Sun J, Manthos G, Schagerl M, Sventzouri E, Al-Tohamy R, Kornaros M, Ali SS (2024) Recent advances in sustainable hydrogen production from microalgae: mechanisms, challenges, and future perspectives. Ecotoxicol Environ Saf 270:115908. https://doi.org/10.1016/j.ecoenv.2023.115908
80. Kaffarov V (1985) Methods of cybernetics in chemistry and chemical technology (in Russian). Moscow
81. Kaffarov V, Vinarov A, Gordeev L (1979) Modelling biochemical reactors (in Russian). Forrest Industry, Moscow
82. Kaffarov V, Vinarov A, Gordeev L (1985) Modelling and system analysis of biochemical industrial production (in Russian). Forrest Industry, Moscow
83. Khoo KS, Ahmad I, Chew KW, Iwamoto K, Bhatnagar A, Show PL (2023) Enhanced microalgal lipid production for biofuel using different strategies including genetic modification of microalgae: a review. Prog Energy Combust Sci 96:101071. https://doi.org/10.1016/j.pecs.2023.101071
84. Kim G-Y, Heo J, Kim K, Chung J, Han J-I (2021) Electrochemical pH control and carbon supply for microalgae cultivation. Chem Eng J 426:131796. https://doi.org/10.1016/j.cej.2021.131796
85. Kim J, Yoo G, Lee H, Lim J, Kim K, Kim CW, Park MS, Yang J-W (2013) Methods of downstream processing for the production of biodiesel from microalgae. Biotechnol Adv 31(6):862–876. https://doi.org/10.1016/j.biotechadv.2013.04.006
86. Kim SH, Sunwoo IY, Hong HJ, Awah CC, Jeong G-T, Kim S-K (2019) Lipid and unsaturated fatty acid productions from three microalgae using nitrate and light-emitting diodes with complementary LED wavelength in a two-phase culture system. Bioprocess Biosyst Eng 42(9):1517–1526. https://doi.org/10.1007/s00449-019-02149-y
87. Kovacevic V, Wesseler J (2010) Cost-effectiveness analysis of algae energy production in the EU. Energy Policy 38(10):5749–5757. https://doi.org/10.1016/j.enpol.2010.05.025
88. Krämer LC, Wasser D, Haitz F, Sabel B, Büchel C (2022) Heterologous expression of HUP1 glucose transporter enables low-light mediated growth on glucose in phaeodactylum tricornutum. Algal Res 64:102719. https://doi.org/10.1016/j.algal.2022.102719

89. Kroumov AD, Campus R, Schreiner R (2002) A dynamic and static analysis of ethanol production industrial process. Tenth Congress of the Bulgarian Microbiologists with International Participation
90. Kroumov AD, Dimitrov D, Gordeev L (1991) Prediction of the functioning of a cell bioreactor model for purification of industrial effluents (in Russian). Int. J. Bioautomation 9:11–18
91. Kroumov AD, Gacheva GV, Iliev II, Alexandrov SD, Pilarski PS, Petkov GD (2013) Analysis of Sf/V ratio of photobioreactors linked with algal physiology. Genetics Plant Physiol 3(1–2)
92. Kroumov AD, Gordeev L (1990) A mathematical model of the biosynthesis process and parametrical sensitivity of the column fermenter (in Russian). Int. J. Bioautomation 5:3–7
93. Kroumov AD, Módenes AN, Tait MCDA (2006) Development of new unstructured model for simultaneous saccharification and fermentation of starch to ethanol by recombinant strain. Biochem Eng J 28(3):243–255. https://doi.org/10.1016/j.bej.2005.11.008
94. Kroumov AD, Módenes AN, Trigueros DEG (2015) A complex theoretical approach for algal medium optimization for CO2 fixation from flue gas. Acta Microbiol Bulg 31(1):61–70
95. Kroumov AD, Módenes AN, Trigueros DEG, Espinoza-Quiñones FR, Borba CE, Scheufele FB, Hinterholz CL (2016) A systems approach for CO_2 fixation from flue gas by microalgae—theory review. Process Biochem 51(11):1817–1832. https://doi.org/10.1016/j.procbio.2016.05.019
96. Kroumov AD, Módenes AN, Wenzel BM (2007) Desenvolvimento de um modelo da cinética enzimática da transesterificação de óleos vegetais para produção de biodiesel. Acta Sci. Technol. 29(1):9–16
97. Kroumov AD, Scheufele FB, Zaharieva MM, Módenes AN, Trigueros DEG, Borba CE, Espinoza-Quiñones FR, Najdenski HM (2021) Complex analysis of new unique human society life in eight coordinate system. Int. J. Bioautomation 25(1):53–72
98. Krzemińska I, Szymańska M, Ciempiel W, Piasecka A (2023) Auxin supplementation under nitrogen limitation enhanced oleic acid and MUFA content in eustigmatos calaminaris biomass with potential for biodiesel production. Sci Rep 13(1):594. https://doi.org/10.1038/s41598-023-27778-y
99. Kumar G, Shekh A, Jakhu S, Sharma Y, Kapoor R, Sharma TR (2020) Bioengineering of microalgae: recent advances, perspectives, and regulatory challenges for industrial application. Front Bioeng Biotechnol 8
100. Kumar V, Nanda M, Kumar S, Chauhan PK (2018) The effects of ultraviolet radiation on growth, biomass, lipid accumulation and biodiesel properties of microalgae. Energy Sources, Part A: Recover, Util, Environ Eff 40(7):787–793. https://doi.org/10.1080/15567036.2018.1463310
101. Kwak HS, Kim JYH, Woo HM, Jin E, Min BK, Sim SJ (2016) Synergistic effect of multiple stress conditions for improving microalgal lipid production. Algal Res 19:215–224. https://doi.org/10.1016/j.algal.2016.09.003
102. Lakshmikandan M, Wang S, Murugesan AG, Saravanakumar M, Selvakumar G (2021) Co-cultivation of Streptomyces and microalgal cells as an efficient system for biodiesel production and bioflocculation formation. Biores Technol 332:125118. https://doi.org/10.1016/j.biortech.2021.125118
103. Lee OK, Seong DH, Lee CG, Lee EY (2015) Sustainable production of liquid biofuels from renewable microalgae biomass. J Ind Eng Chem 29:24–31. https://doi.org/10.1016/j.jiec.2015.04.016
104. Lee S-J, Go S, Jeong G-T, Kim S-K (2011) Oil production from five marine microalgae for the production of biodiesel. Biotechnol Bioprocess Eng 16(3):561–566. https://doi.org/10.1007/s12257-010-0360-0
105. Legodi LM, Moganedi KLM (2024) The potential of lignocellulosic biomass hydrolysates for microbial oil production using yeasts and microalgae. Int J Chem Eng 2024(1):5153495. https://doi.org/10.1155/2024/5153495
106. Legrand J, Artu A, Pruvost J (2021) A review on photobioreactor design and modelling for microalgae production. React Chem and Eng 6. https://doi.org/10.1039/D0RE00450B

107. Leong YK, Su R-H, Varjani S, Lee D-J, Chang J-S (2024) Integrated biohydrogen production and dairy manure wastewater treatment via a microalgae platform. Int J Hydrogen Energy 52:404–417. https://doi.org/10.1016/j.ijhydene.2023.06.308
108. Li D, Yuan Y, Cheng D, Zhao Q (2019) Effect of light quality on growth rate, carbohydrate accumulation, fatty acid profile and lutein biosynthesis of Chlorella sp. AE10. Bioresour Technol 291:121783. https://doi.org/10.1016/j.biortech.2019.121783
109. Li H, Sun X, Sun Y, Ye L, Xue H, Gao F, Yang Y (2024) Enhancing lipid production in Chlorella under successive stresses of periodic micro-current and salinity: performance and contribution. Chem Eng J 486:150409. https://doi.org/10.1016/j.cej.2024.150409
110. Li T, Gargouri M, Feng J, Park J-J, Gao D, Miao C, Dong T, Gang DR, Chen S (2015) Regulation of starch and lipid accumulation in a microalga chlorella sorokiniana. Biores Technol 180:250–257. https://doi.org/10.1016/j.biortech.2015.01.005
111. Li Z, Yang S-T, Zhou Z-Y, Peng S-Y, Zhang Q-H, Long H-Z, Li H-G (2021) Enhancement of lipid production in desmodesmus intermedius Z8 by ultrasonic stimulation coupled with nitrogen and phosphorus stress. Biochem Eng J 172:108061. https://doi.org/10.1016/j.bej.2021.108061
112. Liang S-Y, Li L-H, Ling M-X, Mo L, Huang L-M, Ruan L-R, Shang C-H (2024) Combined effects of nitrogen limitation and overexpression of malic enzyme gene on lipid accumulation in dunaliella parva. Biores Technol 410:131291. https://doi.org/10.1016/j.biortech.2024.131291
113. Liao Q, Sun Y, Huang Y, Xia A, Fu Q, Zhu X (2017) Simultaneous enhancement of Chlorella vulgaris growth and lipid accumulation through the synergy effect between light and nitrate in a planar waveguide flat-plate photobioreactor. Biores Technol 243:528–538. https://doi.org/10.1016/j.biortech.2017.06.091
114. Lin JY, Sri Wahyu Effendi S, Ng IS (2022) Enhanced carbon capture and utilization (CCU) using heterologous carbonic anhydrase in chlamydomonas reinhardtii for lutein and lipid production. Bioresour Technol 351:127009. https://doi.org/10.1016/j.biortech.2022.127009
115. Lin W-R, Lai Y-C, Sung P-K, Tan S-I, Chang C-H, Chen C-Y, Chang J-S, Ng IS (2018) Enhancing carbon capture and lipid accumulation by genetic carbonic anhydrase in microalgae. J Taiwan Inst Chem Eng 93:131–141. https://doi.org/10.1016/j.jtice.2018.10.010
116. Liu J, Mao X, Zhou W, Guarnieri MT (2016) Simultaneous production of triacylglycerol and high-value carotenoids by the astaxanthin-producing oleaginous green microalga chlorella zofingiensis. Biores Technol 214:319–327. https://doi.org/10.1016/j.biortech.2016.04.112
117. Liu T, Li Y, Liu F, Wang C (2016) The enhanced lipid accumulation in oleaginous microalga by the potential continuous nitrogen-limitation (CNL) strategy. Biores Technol 203:150–159. https://doi.org/10.1016/j.biortech.2015.12.021
118. Lower SK (1999) Carbonate Equilibria in Natural Waters. Simon Fraser University
119. Lu D, Tabil LG, Wang D, Wang G, Emami S (2014) Experimental trials to make wheat straw pellets with wood residue and binders. Biomass Bioenerg 69:287–296. https://doi.org/10.1016/j.biombioe.2014.07.029
120. Lynch F, Santana-Sánchez A, Jämsä M, Sivonen K, Aro E-M, Allahverdiyeva Y (2015) Screening native isolates of cyanobacteria and a green alga for integrated wastewater treatment, biomass accumulation and neutral lipid production. Algal Res 11:411–420. https://doi.org/10.1016/j.algal.2015.05.015
121. Ma C, Zhang Y-B, Ho S-H, Xing D-F, Ren N-Q, Liu B-F (2017) Cell growth and lipid accumulation of a microalgal mutant Scenedesmus sp. Z-4 by combining light/dark cycle with temperature variation. Biotechnol Biofuels 10(1):260. https://doi.org/10.1186/s13068-017-0948-0
122. Ma R, Thomas-Hall SR, Chua ET, Eltanahy E, Netzel ME, Netzel G, Lu Y, Schenk PM (2018) LED power efficiency of biomass, fatty acid, and carotenoid production in Nannochloropsis microalgae. Biores Technol 252:118–126. https://doi.org/10.1016/j.biortech.2017.12.096
123. Ma S, Huang Y, Zhu X, Xia A, Zhu X, Liao Q (2024) Growth-based dynamic light transmission modeling and optimization in microalgal photobioreactors for high efficiency CO_2 fixation. Renew Sustain Energy Rev 197:114414. https://doi.org/10.1016/j.rser.2024.114414

124. Ma Y, Wang Z, Yu C, Yin Y, Zhou G (2014) Evaluation of the potential of 9 Nannochloropsis strains for biodiesel production. Biores Technol 167:503–509. https://doi.org/10.1016/j.biortech.2014.06.047
125. Maeda Y, Yoshino T, Matsunaga T, Matsumoto M, Tanaka T (2018) Marine microalgae for production of biofuels and chemicals. Curr Opin Biotechnol 50:111–120. https://doi.org/10.1016/j.copbio.2017.11.018
126. Maltsev Y, Krivova Z, Maltseva S, Maltseva K, Gorshkova E, Kulikovskiy M (2021) Lipid accumulation by coelastrella multistriata (scenedesmaceae, sphaeropleales) during nitrogen and phosphorus starvation. Sci Rep 11(1):19818. https://doi.org/10.1038/s41598-021-99376-9
127. Maltsev Y, Maltseva I, Maltseva S, Kociolek JP, Kulikovskiy M (2021) A new species of freshwater algae nephrochlamys yushanlensis sp. nov. (Selenastraceae, sphaeropleales) and its lipid accumulation during nitrogen and phosphorus starvation. J Phycol 57(2):606–618. https://doi.org/10.1111/jpy.13116
128. Mao X, Zhang Y, Wang X, Liu J (2020) Novel insights into salinity-induced lipogenesis and carotenogenesis in the oleaginous astaxanthin-producing alga chromochloris zofingiensis: a multi-omics study. Biotechnol Biofuels 13(1):73. https://doi.org/10.1186/s13068-020-01714-y
129. Marey RS, Abo-Shady AM, Khairy HM, Abd El-Moneim AM, Abomohra A (2024) Enhanced lipid production and essential ω-fatty acids synthesis by the hypersaline biodiesel-promising microalga Tetraselmis elliptica through growth medium optimization. Biomass Conversion and Biorefinery 14(11):12455–12468. https://doi.org/10.1007/s13399-022-03290-7
130. Marin P, Borba CE, Módenes AN, Espinoza-Quiñones FR, de Oliveira SPD, Kroumov AD (2014) Determination of the mass transfer limiting step of dye adsorption onto commercial adsorbent by using mathematical models. Environ Technol 35(18):2356–2364. https://doi.org/10.1080/09593330.2014.904445
131. Mat Husin MA, Mohd Yasin NH, Takriff MS, Jamar NH (2024) A review on pretreatment methods for lipid extraction from microalgae biomass. Prep Biochem Biotechnol 54(2):159–174. https://doi.org/10.1080/10826068.2023.2214923
132. Mata TM, Martins AA, Caetano NS (2010) Microalgae for biodiesel production and other applications: a review. Renew Sustain Energy Rev 14(1):217–232. https://doi.org/10.1016/j.rser.2009.07.020
133. Minhas AK, Hodgson P, Barrow CJ, Adholeya A (2016) A review on the assessment of stress conditions for simultaneous production of microalgal lipids and carotenoids. Front Microbiol 7
134. Módenes AN, Scheufele FB, Espinoza-Quiñones FR, de Souza PSC, Cripa CRB, dos Santos J, Steffen V, Kroumov AD (2015) Adsorption of direct of yellow ARLE dye by activated carbon of shell of coconut palm: diffusional effects on kinetics and equilibrium states. Int. J. Bioautomation 19(2):187–206
135. Moheimani NR (2013) Inorganic carbon and pH effect on growth and lipid productivity of tetraselmis suecica and chlorella sp (Chlorophyta) grown outdoors in bag photobioreactors. J Appl Phycol 25(2):387–398. https://doi.org/10.1007/s10811-012-9873-6
136. Monte Blanco SPD, Scheufele FB, Módenes AN, Espinoza-Quiñones FR, Marin P, Kroumov AD, Borba CE (2017) Kinetic, equilibrium and thermodynamic phenomenological modeling of reactive dye adsorption onto polymeric adsorbent. Chem Eng J 307:466–475. https://doi.org/10.1016/j.cej.2016.08.104
137. Morales M, Aflalo C, Bernard O (2021) Microalgal lipids: a review of lipids potential and quantification for 95 phytoplankton species. Biomass Bioenerg 150:106108. https://doi.org/10.1016/j.biombioe.2021.106108
138. Munisamy Sambasivam K, Kuppan P, Shashirekha V, Tamilarasan K, Abinandan S (2023) Cascading utilization of residual microalgal biomass: Sustainable strategies for energy, environmental and value-added product applications. Bioresource Technology Reports 23:101588. https://doi.org/10.1016/j.biteb.2023.101588

139. Muñoz CF, Südfeld C, Naduthodi MIS, Weusthuis RA, Barbosa MJ, Wijffels RH, D'Adamo S (2021) Genetic engineering of microalgae for enhanced lipid production. Biotechnol Adv 52:107836. https://doi.org/10.1016/j.biotechadv.2021.107836
140. Nascimento IA, Marques SSI, Cabanelas ITD, Pereira SA, Druzian JI, de Souza CO, Vich DV, de Carvalho GC, Nascimento MA (2013) Screening microalgae strains for biodiesel production: lipid productivity and estimation of fuel quality based on fatty acids profiles as selective criteria. BioEnergy Research 6(1):1–13. https://doi.org/10.1007/s12155-012-9222-2
141. Munir N, Imtiaz A, Nadia, Sharif N, Shagufta, Naz S (2015) Optimization of growth conditions of different algal strains and determination of their lipid contents. J Anim Plant Sci 25:546–553
142. Negi S, Barry AN, Friedland N, Sudasinghe N, Subramanian S, Pieris S, Holguin FO, Dungan B, Schaub T, Sayre R (2016) Impact of nitrogen limitation on biomass, photosynthesis, and lipid accumulation in chlorella sorokiniana. J Appl Phycol 28(2):803–812. https://doi.org/10.1007/s10811-015-0652-z
143. Nicodemou A, Konstantinou D, Koutinas M (2024) Enhanced biomass and lipid production from olive processing wastewater using scenedesmus obliquus in a two-stage cultivation strategy under salt stress. Biochem Eng J 205:109290. https://doi.org/10.1016/j.bej.2024.109290
144. Osman MEH, Abo-Shady AM, Gheda SF, Desoki SM, Elshobary ME (2023) Unlocking the potential of microalgae cultivated on wastewater combined with salinity stress to improve biodiesel production. Environ Sci Pollut Res 30(53):114610–114624. https://doi.org/10.1007/s11356-023-30370-6
145. Pandey S, Narayanan I, Selvaraj R, Varadavenkatesan T, Vinayagam R (2024) Biodiesel production from microalgae: a comprehensive review on influential factors, transesterification processes, and challenges. Fuel 367:131547. https://doi.org/10.1016/j.fuel.2024.131547
146. Pandit PR, Fulekar MH, Karuna MSL (2017) Effect of salinity stress on growth, lipid productivity, fatty acid composition, and biodiesel properties in acutodesmus obliquus and chlorella vulgaris. Environ Sci Pollut Res 24(15):13437–13451. https://doi.org/10.1007/s11356-017-8875-y
147. Parichehreh R, Gheshlaghi R, Mahdavi MA, Kamyab H (2021) Investigating the effects of eleven key physicochemical factors on growth and lipid accumulation of chlorella sp. as a feedstock for biodiesel production. J Biotechnol 340:64–74. https://doi.org/10.1016/j.jbiotec.2021.08.010
148. Peng X, Liu S, Zhang W, Zhao Y, Chen L, Wang H, Liu T (2014) Triacylglycerol accumulation of phaeodactylum tricornutum with different supply of inorganic carbon. J Appl Phycol 26(1):131–139. https://doi.org/10.1007/s10811-013-0075-7
149. Perez-Garcia O, Escalante FME, de-Bashan LE, Bashan Y (2011) Heterotrophic cultures of microalgae: metabolism and potential products. Water Research 45(1):11–36. https://doi.org/10.1016/j.watres.2010.08.037
150. Perin G, Bellan A, Bernardi A, Bezzo F, Morosinotto T (2019) The potential of quantitative models to improve microalgae photosynthetic efficiency. Physiol Plant 166(1):380–391. https://doi.org/10.1111/ppl.12915
151. Plöhn M, Spain O, Sirin S, Silva M, Escudero-Oñate C, Ferrando-Climent L, Allahverdiyeva Y, Funk C (2021) Wastewater treatment by microalgae. Physiol Plant 173(2):568–578. https://doi.org/10.1111/ppl.13427
152. Popovich CA, Damiani C, Constenla D, Martínez AM, Freije H, Giovanardi M, Pancaldi S, Leonardi PI (2012) Neochloris oleoabundans grown in enriched natural seawater for biodiesel feedstock: evaluation of its growth and biochemical composition. Biores Technol 114:287–293. https://doi.org/10.1016/j.biortech.2012.02.121
153. Popovich CA, Pistonesi M, Hegel P, Constenla D, Bielsa GB, Martín LA, Damiani MC, Leonardi PI (2019) Unconventional alternative biofuels: quality assessment of biodiesel and its blends from marine diatom navicula cincta. Algal Res 39:101438. https://doi.org/10.1016/j.algal.2019.101438
154. Praveenkumar R, Shameera K, Mahalakshmi G, Akbarsha MA, Thajuddin N (2012) Influence of nutrient deprivations on lipid accumulation in a dominant indigenous microalga chlorella

sp., BUM11008: evaluation for biodiesel production. Biomass Bioenerg 37:60–66. https://doi.org/10.1016/j.biombioe.2011.12.035
155. Proietti Tocca G, Agostino V, Menin B, Tommasi T, Fino D, Di Caprio F (2024) Mixotrophic and heterotrophic growth of microalgae using acetate from different production processes. Rev Environ Sci Bio/Technol 23(1):93–132. https://doi.org/10.1007/s11157-024-09682-7
156. Pruvost J, Van Vooren G, Cogne G, Legrand J (2009) Investigation of biomass and lipids production with neochloris oleoabundans in photobioreactor. Biores Technol 100(23):5988–5995. https://doi.org/10.1016/j.biortech.2009.06.004
157. Qiu R, Gao S, Lopez PA, Ogden KL (2017) Effects of pH on cell growth, lipid production and CO_2 addition of microalgae chlorella sorokiniana. Algal Res 28:192–199. https://doi.org/10.1016/j.algal.2017.11.004
158. Rady HA, Ali SS, El-Sheekh MM (2024) Strategies to enhance biohydrogen production from microalgae: a comprehensive review. J Environ Manage 356:120611. https://doi.org/10.1016/j.jenvman.2024.120611
159. Rai MP, Gupta S (2017) Effect of media composition and light supply on biomass, lipid content and FAME profile for quality biofuel production from scenedesmus abundans. Energy Convers Manage 141:85–92. https://doi.org/10.1016/j.enconman.2016.05.018
160. Ramachandra TV, Hebbale D (2020) Bioethanol from macroalgae: prospects and challenges. Renew Sustain Energy Rev 117:109479. https://doi.org/10.1016/j.rser.2019.109479
161. Rastogi RP, Pandey A, Larroche C, Madamwar D (2018) Algal Green Energy–R&D and technological perspectives for biodiesel production. Renew Sustain Energy Rev 82:2946–2969. https://doi.org/10.1016/j.rser.2017.10.038
162. Ratomski P, Hawrot-Paw M (2021) Influence of nutrient-stress conditions on chlorella vulgaris biomass production and lipid content. Catalysts 11. https://doi.org/10.3390/catal11050573
163. Rawindran H, Alam MM, Sahrin NT, Raksasat R, Leong WH, Liew CS, Supramaniam U, Lim JW, Usman A, Tong W-Y, Suresh S, Khoo KS (2024) Recent advancements in harnessing biodiesel from microalgae through attached growth systems. Biocatal Agric Biotechnol 58:103205. https://doi.org/10.1016/j.bcab.2024.103205
164. Ren H-Y, Liu B-F, Kong F, Zhao L, Xie G-J, Ren N-Q (2014) Enhanced lipid accumulation of green microalga scenedesmus sp. by metal ions and EDTA addition. Biores Technol 169:763–767. https://doi.org/10.1016/j.biortech.2014.06.062
165. Ren H-Y, Zhu J-N, Kong F, Xing D, Zhao L, Ma J, Ren N-Q, Liu B-F (2019) Ultrasonic enhanced simultaneous algal lipid production and nutrients removal from non-sterile domestic wastewater. Energy Convers Manage 180:680–688. https://doi.org/10.1016/j.enconman.2018.11.028
166. Reyimu Z, Özçimen D (2017) Batch cultivation of marine microalgae nannochloropsis oculata and tetraselmis suecica in treated municipal wastewater toward bioethanol production. J Clean Prod 150:40–46. https://doi.org/10.1016/j.jclepro.2017.02.189
167. Richmond A Ed (2004) Handbook of microalgal culture: biotechnology and applied phycology. Wiley-Blackwell
168. Roels JA Ed (1983) Energetics and kinetics in biotechnology. Elsevier Biomedical Press
169. Ruangsomboon S (2012) Effect of light, nutrient, cultivation time and salinity on lipid production of newly isolated strain of the green microalga, Botryococcus braunii KMITL 2. Biores Technol 109:261–265. https://doi.org/10.1016/j.biortech.2011.07.025
170. Saini R, Osorio-Gonzalez CS, Brar SK, Kwong R (2021) A critical insight into the development, regulation and future prospects of biofuels in Canada. Bioengineered 12(2):9847–9859. https://doi.org/10.1080/21655979.2021.1996017
171. Salama E-S, Jeon B-H, Chang SW, Lee S-H, Roh H-S, Yang I-S, Kurade MB, El-Dalatony MM, Kim D-H, Kim K-H, Kim S (2017) Interactive effect of indole-3-acetic acid and diethyl aminoethyl hexanoate on the growth and fatty acid content of some microalgae for biodiesel production. J Clean Prod 168:1017–1024. https://doi.org/10.1016/j.jclepro.2017.09.057
172. Sánchez-Zurano A, Villaró-Cos S, Ciardi M, Acién-Fernández FG, Fernández-Sevilla JM, Lafarga T (2024) Assessment of the mixotrophic production of chlorella vulgaris using milk whey as a nutrient source. J Appl Phycol 36(1):87–100. https://doi.org/10.1007/s10811-023-03142-5

173. Sánchez JF, Fernández JM, Acién FG, Rueda A, Pérez-Parra J, Molina E (2008) Influence of culture conditions on the productivity and lutein content of the new strain scenedesmus almeriensis. Process Biochem 43(4):398–405. https://doi.org/10.1016/j.procbio.2008.01.004
174. Saravanan A, Deivayanai VC, Senthil Kumar P, Rangasamy G, Varjani S (2022) CO_2 biomitigation using genetically modified algae and biofuel production towards a carbon net-zero society. Biores Technol 363:127982. https://doi.org/10.1016/j.biortech.2022.127982
175. Scheufele FB, Hinterholz CL, Zaharieva MM, Najdenski HM, Módenes AN, Trigueros DEG, Borba CE, Espinoza-Quiñones FR, Kroumov AD (2019) Complex mathematical analysis of photobioreactor system. Eng Life Sci 19(12):844–859. https://doi.org/10.1002/elsc.201800044
176. Schuelter AR, Kroumov AD, Hinterholz CL, Fiorini A, Trigueros DEG, Vendruscolo EG, Zaharieva MM, Módenes AN (2019) Isolation and identification of new microalgae strains with antibacterial activity on food-borne pathogens. Engineering approach to optimize synthesis of desired metabolites. Biochem Eng J 144:28–39. https://doi.org/10.1016/j.bej.2019.01.007
177. Shrestha N, Dandinpet KK, Schneegurt MA (2020) Effects of nitrogen and phosphorus limitation on lipid accumulation by chlorella kessleri str. UTEX 263 grown in darkness. J Appl Phycol 32(5):2795–2805. https://doi.org/10.1007/s10811-020-02144-x
178. Shuba ES, Kifle D (2018) Microalgae to biofuels: 'Promising' alternative and renewable energy, review. Renew Sustain Energy Rev 81:743–755. https://doi.org/10.1016/j.rser.2017.08.042
179. Silva-Gálvez AL, López-Sánchez A, Camargo-Valero MA, Prosenc F, González-López ME, Gradilla-Hernández MS (2024) Strategies for livestock wastewater treatment and optimised nutrient recovery using microalgal-based technologies. J Environ Manage 354:120258. https://doi.org/10.1016/j.jenvman.2024.120258
180. Singh J, Jain D, Agarwal P, Singh RP (2020) Auxin and cytokinin synergism augmenting biomass and lipid production in microalgae desmodesmus sp. JS07. Process Biochem 95:223–234. https://doi.org/10.1016/j.procbio.2020.02.012
181. Singh NK, Dhar DW (2011) Microalgae as second generation biofuel. A review. Agron Sustain Dev 31(4):605–629. https://doi.org/10.1007/s13593-011-0018-0
182. Singh RP, Yadav P, Kumar A, Hashem A, Al-Arjani ABF, Abd_Allah EF, Rodríguez Dorantes A, Gupta RK (2022) Physiological and biochemical responses of bicarbonate supplementation on biomass and lipid content of green algae Scenedesmus sp. BHU1 isolated from wastewater for renewable biofuel feedstock. Frontiers Microbiology 13
183. Singh RP, Yadav P, Kumar A, Hashem A, Avila-Quezada GD, Abd Allah EF, Gupta RK (2023) Salinity-induced physiochemical alterations to enhance lipid content in oleaginous microalgae Scenedesmus sp. BHU1 via Two-Stage Cultivation for Biodiesel Feedstock. Microorganisms 11. https://doi.org/10.3390/microorganisms11082064
184. Sirajunnisa AR, Surendhiran D (2016) Algae–a quintessential and positive resource of bioethanol production: a comprehensive review. Renew Sustain Energy Rev 66:248–267. https://doi.org/10.1016/j.rser.2016.07.024
185. Snow AA, Smith VH (2012) Genetically engineered algae for biofuels: a key role for ecologists. Bioscience 62(8):765–768. https://doi.org/10.1525/bio.2012.62.8.9
186. Song X, Liu B-F, Kong F, Ren N-Q, Ren H-Y (2022) Overview on stress-induced strategies for enhanced microalgae lipid production: application, mechanisms and challenges. Resour Conserv Recycl 183:106355. https://doi.org/10.1016/j.resconrec.2022.106355
187. Song X, Zhao Y, Han B, Li T, Zhao P, Xu JW, Yu X (2020) Strigolactone mediates jasmonic acid-induced lipid production in microalga monoraphidium sp. QLY-1 under nitrogen deficiency conditions. Bioresource Technol 306:123107. https://doi.org/10.1016/j.biortech.2020.123107
188. Song X, Zhao Y, Li T, Han B, Zhao P, Xu JW, Yu X (2019) Enhancement of lipid accumulation in monoraphidium sp. QLY-1 by induction of strigolactone. Bioresour Technol 288:121607. https://doi.org/10.1016/j.biortech.2019.121607

189. Sousa V, Pereira RN, Vicente AA, Dias O, Geada P (2023) Microalgae biomass as an alternative source of biocompounds: new insights and future perspectives of extraction methodologies. Food Res Int 173:113282. https://doi.org/10.1016/j.foodres.2023.113282
190. Spalding MH, Critchley C, Govindjee, Orgren WL (1984) Influence of carbon dioxide concentration during growth on fluorescence induction characteristics of the Green Alga Chlamydomonas reinhardii. Photosynth Res 5(2):169–176. https://doi.org/10.1007/BF00028529
191. Srinivasan R, Mageswari A, Subramanian P, Suganthi C, Chaitanyakumar A, Aswini V, Gothandam KM (2018) Bicarbonate supplementation enhances growth and biochemical composition of dunaliella salina V-101 by reducing oxidative stress induced during macronutrient deficit conditions. Sci Rep 8(1):6972. https://doi.org/10.1038/s41598-018-25417-5
192. Srivastava G, Nishchal, Goud VV (2017) Salinity induced lipid production in microalgae and cluster analysis (ICCB 16-BR_047). Bioresource Techn 242:244–252. https://doi.org/10.1016/j.biortech.2017.03.175
193. Suastes-Rivas JK, Hernández-Altamirano R, Mena-Cervantes VY, Chairez I (2020) Simultaneous optimization of biomass and metabolite production by a microalgae-yeast co-culture under inorganic micronutrients. BioEnergy Research 13(3):974–985. https://doi.org/10.1007/s12155-020-10116-9
194. Subramani L, Venu H (2019) Evaluation of methyl ester derived from novel chlorella emersonii as an alternative feedstock for DI diesel engine & its combustion, performance and tailpipe emissions. Heat Mass Transf 55(5):1513–1534. https://doi.org/10.1007/s00231-018-2530-0
195. Sun XM, Ren LJ, Zhao QY, Ji XJ, Huang H (2019) Enhancement of lipid accumulation in microalgae by metabolic engineering. Biochimica et Biophysica Acta (BBA)–Molecular and Cell Biology of Lipids 1864(4):552–566. https://doi.org/10.1016/j.bbalip.2018.10.004
196. Suparmaniam U, Lam MK, Lim JW, Yusup S, Tan IS, Lau SY, Kodgire P, Kachhwaha SS (2023) Influence of environmental stress on microalgae growth and lipid profile: a systematic review. Phytochem Rev 22(4):879–901. https://doi.org/10.1007/s11101-022-09810-7
197. Talhami M, Mussa AA, Thaher MI, Das P, Abouelela AR, Hawari AH (2023) Efficient extraction of lipids from microalgal biomass for the production of biofuels using low-cost protic ionic solvents. Biochem Eng J 194:108902. https://doi.org/10.1016/j.bej.2023.108902
198. Tan S, Wen F, Liu D, Lu H, Li L, Zhu L (2024) Physiological responses and lipid accumulation of freshwater microalgae chlorella sorokiniana under short-term zinc stress in water solution. Algal Res 80:103528. https://doi.org/10.1016/j.algal.2024.103528
199. Tan X-B, Yang L-B, Zhang W-W, Zhao X-C (2020) Lipids production and nutrients recycling by microalgae mixotrophic culture in anaerobic digestate of sludge using wasted organics as carbon source. Biores Technol 297:122379. https://doi.org/10.1016/j.biortech.2019.122379
200. Tang DYY, Chew KW, Gentili FG, Wang C, Munawaroh HSH, Ma Z, Sun F, Govarthanan M, Alharthi S, Show PL (2023) Dechlorophyllization of microalgae biomass for the bioconversion into lipid-rich bioproducts. Ind Eng Chem Res 62(36):14478–14483. https://doi.org/10.1021/acs.iecr.3c00780
201. Tourang M, Baghdadi M, Torang A, Sarkhosh S (2019) Optimization of carbohydrate productivity of spirulina microalgae as a potential feedstock for bioethanol production. Int J Environ Sci Technol 16(3):1303–1318. https://doi.org/10.1007/s13762-017-1592-8
202. Trigueros DEG, Hinterholz CL, Fiorese ML, Aragão GMF, Schmidell W, Reis MAM, Kroumov AD (2017) Statistical evaluation and discrimination of competing kinetic models and hypothesis for the mathematical description of poly-3(hydroxybutyrate) synthesis by Cupriavidus necator DSM 545. Chem Eng Sci 160:20–33. https://doi.org/10.1016/j.ces.2016.11.025
203. Trigueros DEG, Módenes AN, Espinoza-Quiñones FR, Kroumov AD (2010) The evaluation of benzene and phenol biodegradation kinetics by applying non-structured models. Water Sci Technol 61(5):1289–1298. https://doi.org/10.2166/wst.2010.034
204. Trigueros DEG, Modenes AN, Kroumov AD (2007) Modeling biodegradation kinetics on benzene and toluene and their mixture. Bioautomation 7:9–22

205. Trigueros DEG, Módenes AN, Kroumov AD, Espinoza-Quiñones FR (2010) Modeling of biodegradation process of BTEX compounds: kinetic parameters estimation by using particle swarm global optimizer. Process Biochem 45(8):1355–1361. https://doi.org/10.1016/j.procbio.2010.05.007
206. Tripathi R, Singh J, Thakur IS (2015) Characterization of microalga Scenedesmus sp. ISTGA1 for potential CO_2 sequestration and biodiesel production. Renewable Energy 74:774–781. https://doi.org/10.1016/j.renene.2014.09.005
207. Tu Z, Liu L, Lin W, Xie Z, Luo J (2018) Potential of using sodium bicarbonate as external carbon source to cultivate microalga in non-sterile condition. Biores Technol 266:109–115. https://doi.org/10.1016/j.biortech.2018.06.076
208. Tüccar G, Aydın K (2013) Evaluation of methyl ester of microalgae oil as fuel in a diesel engine. Fuel 112:203–207. https://doi.org/10.1016/j.fuel.2013.05.016
209. Udayan A, Sabapathy H, Arumugam M (2020) Stress hormones mediated lipid accumulation and modulation of specific fatty acids in Nannochloropsis oceanica CASA CC201. Biores Technol 310:123437. https://doi.org/10.1016/j.biortech.2020.123437
210. Valdés FJ, Hernández MR, Catalá L, Marcilla A (2012) Estimation of CO_2 stripping/CO_2 microalgae consumption ratios in a bubble column photobioreactor using the analysis of the pH profiles. Application to Nannochloropsis oculata microalgae culture. Biores Technol 119:1–6. https://doi.org/10.1016/j.biortech.2012.05.120
211. Vendruscolo RG, Fagundes MB, Maroneze MM, do Nascimento TC, de Menezes CR, Barin JS, Zepka LQ, Jacob-Lopes E, Wagner R (2019) Scenedesmus obliquus metabolomics: effect of photoperiods and cell growth phases. Bioprocess Biosyst Eng 42(5):727–739. https://doi.org/10.1007/s00449-019-02076-y
212. Wang H, Fu R, Pei G (2012) A study on lipid production of the mixotrophic microalgae Phaeodactylum tricornutum on various carbon sources. Afr J Microbiol Res 6. https://doi.org/10.5897/AJMR11.1365.
213. Wang L, Min M, Li Y, Chen P, Chen Y, Liu Y, Wang Y, Ruan R (2010) Cultivation of green algae chlorella sp. in different wastewaters from municipal wastewater treatment plant. Appl Biochem Biotechnol 162(4):1174–1186. https://doi.org/10.1007/s12010-009-8866-7
214. Wang L, Nancollas GH (2008) Calcium orthophosphates: crystallization and dissolution. Chem Rev 108(11):4628–4669. https://doi.org/10.1021/cr0782574
215. Wang M, Liu H, Qiao K, Ye X, Takano T, Liu S, Bu Y (2021) Exogenous $NaHCO_3$ enhances growth and lipid accumulation of the highly $NaHCO_3$-tolerant Nannochloris sp. JB17. J Appl Phycol 33(1):241–253. https://doi.org/10.1007/s10811-020-02293-z
216. Wei L, Huang X, Huang Z (2015) Temperature effects on lipid properties of microalgae tetraselmis subcordiformis and nannochloropsis oculata as biofuel resources. Chin J Oceanol Limnol 33(1):99–106. https://doi.org/10.1007/s00343-015-3346-0
217. Wei Z, Wang H, Li X, Zhao Q, Yin Y, Xi L, Ge B, Qin S (2020) Enhanced biomass and lipid production by co-cultivation of chlorella vulgaris with mesorhizobium sangaii under nitrogen limitation. J Appl Phycol 32(1):233–242. https://doi.org/10.1007/s10811-019-01924-4
218. Wenzel B, Tait M, Modenes A, Kroumov A (2006) Modelling chemical kinetics of soybean oil transesterification process for biodiesel production: an analysis of molar ratio between alcohol and soybean oil temperature changes on the process conversion rate. Bioautomation 5:13–22
219. Wenzel BM, Tait M, Módenes AN, Kroumov AD (2006b) Development of a novel enzymatic kinetic model of oils transesterification and esterification for biodiesel production. In: 19th international symposium-bioprocess systems, Sofia, Bulgaria
220. Wijffels RH, Barbosa MJ, Eppink MHM (2010) Microalgae for the production of bulk chemicals and biofuels. Biofuels, Bioprod Biorefin 4(3):287–295. https://doi.org/10.1002/bbb.215
221. Wood JM, Wang H-K (1983) Microbial resistance to heavy metals. Environ Sci Technol 17(12):582A-590A. https://doi.org/10.1021/es00118a717
222. Wu H, Miao X (2014) Biodiesel quality and biochemical changes of microalgae chlorella pyrenoidosa and scenedesmus obliquus in response to nitrate levels. Biores Technol 170:421–427. https://doi.org/10.1016/j.biortech.2014.08.017

223. Wu Y-H, Yu Y, Hu H-Y, Zhuang L-L (2016) Effects of cultivation conditions on the production of soluble algal products (SAPs) of scenedesmus sp. LX1. Algal Res 16:376–382. https://doi.org/10.1016/j.algal.2016.04.006
224. Xie Z, Lin W, Luo J (2021) Co-cultivation of microalga and xylanolytic bacterium by a continuous two-step strategy to enhance algal lipid production. Biores Technol 330:124953. https://doi.org/10.1016/j.biortech.2021.124953
225. Xin L, Hong-ying H, Jia Y (2010) Lipid accumulation and nutrient removal properties of a newly isolated freshwater microalga, scenedesmus sp. LX1, growing in secondary effluent. New Biotechnol 27(1):59–63. https://doi.org/10.1016/j.nbt.2009.11.006
226. Xu D, Gao Z, Li F, Fan X, Zhang X, Ye N, Mou S, Liang C, Li D (2013) Detection and quantitation of lipid in the microalga tetraselmis subcordiformis (Wille) butcher with BODIPY 505/515 staining. Biores Technol 127:386–390. https://doi.org/10.1016/j.biortech.2012.09.068
227. Yang L, Chen J, Qin S, Zeng M, Jiang Y, Hu L, Xiao P, Hao W, Hu Z, Lei A, Wang J (2018) Growth and lipid accumulation by different nutrients in the microalga chlamydomonas reinhardtii. Biotechnol Biofuels 11(1):40. https://doi.org/10.1186/s13068-018-1041-z
228. You Z, Zhang Q, Peng Z, Miao X (2019) Lipid droplets mediate salt stress tolerance in parachlorella kessleri 1. Plant Physiol 181(2):510–526. https://doi.org/10.1104/pp.19.00666
229. Yu H, Kim J, Lee C (2019) Nutrient removal and microalgal biomass production from different anaerobic digestion effluents with chlorella species. Sci Rep 9(1):6123. https://doi.org/10.1038/s41598-019-42521-2
230. Yu N, Dieu LTJ, Harvey S, Lee D-Y (2015) Optimization of process configuration and strain selection for microalgae-based biodiesel production. Biores Technol 193:25–34. https://doi.org/10.1016/j.biortech.2015.05.101
231. Zaharieva MM, Zheleva-Dimitrova D, Rusinova-Videva S, Ilieva Y, Brachkova A, Balabanova V, Gevrenova R, Kim TC, Kaleva M, Georgieva A, Mileva M, Yoncheva K, Benbassat N, Najdenski H, Kroumov AD (2022) Antimicrobial and antioxidant potential of scenedesmus obliquus microalgae in the context of integral biorefinery concept. Molecules 27. https://doi.org/10.3390/molecules27020519
232. Zhang P, Xin Y, He Y, Tang X, Shen C, Wang Q, Lv N, Li Y, Hu Q, Xu J (2022) Exploring a blue-light-sensing transcription factor to double the peak productivity of oil in Nannochloropsis oceanica. Nat Commun 13(1):1664. https://doi.org/10.1038/s41467-022-29337-x
233. Zhao Y, Song X, Zhao P, Li T, Xu JW, Yu X (2021) Role of melatonin in regulation of lipid accumulation, autophagy and salinity-induced oxidative stress in microalga Monoraphidium sp. QLY-1. Algal Research 54:102196. https://doi.org/10.1016/j.algal.2021.102196.
234. Zhou H, Wang J, Zhang Z, Lan CQ (2024) High cell density culture of microalgae in horizontal thin-layer algal reactor: modeling of light attenuation and cell growth kinetics. Chem Eng J 496:154175. https://doi.org/10.1016/j.cej.2024.154175
235. Zhu L (2015) Microalgal culture strategies for biofuel production: a review. Biofuels, Bioprod Biorefin 9(6):801–814. https://doi.org/10.1002/bbb.1576
236. Zhu S, Gu D, Lu C, Zhang C, Chen J, Yang R, Luo Q, Wang T, Zhang P, Chen H (2022) Cold stress tolerance of the intertidal red alga Neoporphyra haitanensis. BMC Plant Biol 22(1):114. https://doi.org/10.1186/s12870-022-03507-x

The Role of Microalgae in Achieving Sustainable Development Goal 6: Clean Water and Sanitation

E. O. Babatunde, T. M. Busari, J. A. Adeyoola, and F. B. Owoyale

Abstract There is a need to sustain microalgae wastewater treatment systems. These systems are evaluated for their technological, operational, environmental, and economic challenges in large-scale microalgae processes. Microalgae reduce greenhouse gas (GHG) emissions and nutrient recovery, though challenges abound. Technological and economic challenges include land use limits, ideal growth conditions, potential contamination from hazardous bacteria and invasive species, and high operational expenses for CO_2 extraction. Despite these challenges, microalgae have substantial prospects for improving water quality and furthering clean water and sanitation. This review emphasizes the significance of conducting complete life-cycle assessments, economic evaluations, and strategic management methods to maximize the viability of microalgae-based wastewater treatment systems and their potential contribution to global sustainability.

Keywords Sustainable development · Clean water · Microalgae · Sanitation

1 Introduction

Water is vital to human health, sustainable communities and ecosystems. 2 billion people worldwide cannot access safe drinking water, while 3 billion people cannot access basic sanitation facilities [1]. Access to safe drinking water and adequate sanitation facilities is vital to public health, and sustainable environment [1]. Sustainable Development Goal 6 (SDG 6) is a global initiative to ensure available and sustainable water management for all by 2030. This goal emphasizes the importance of clean

E. O. Babatunde (✉)
Department of Chemical Engineering Technology, Doornfontein Campus, Faculty of Engineering and Built Environment, University of Johannesburg, Johannesburg, South Africa
e-mail: babatunde.eo@unilorin.edu.ng

E. O. Babatunde · T. M. Busari · J. A. Adeyoola · F. B. Owoyale
Department of Chemical Engineering, Faculty of Engineering and Technology, University of Ilorin, Ilorin, Nigeria

A. Khan and K. A. Alzahrani (eds.), *Microalgae Biotechnology and Sustainable Development*, Sustainable Materials and Technology,
https://doi.org/10.1007/978-981-95-2597-3_5

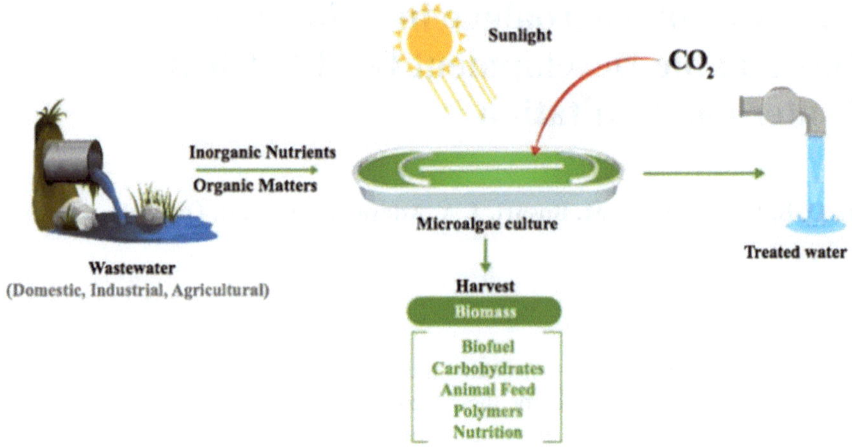

Fig. 1 Microbial wastewater treatment and nutrient recovery

water and sanitation as fundamental human rights. Clean water and sanitation are essential components for achieving health, well-being, and sustainable development.

Under SDG 6, the targets encompass universal access to safe and affordable drinking water. Others are adequate and equitable sanitation and hygiene, and the protection and restoration of water-related ecosystems [2] (Fig. 1). These targets connect water and sanitation with health, education, and gender equality. Accessing clean water and sanitation facilities in schools promotes hygiene and reduces the incidence of waterborne diseases among children. These factors impact children's educational outcomes [3]. Moreover, lack of sanitation facilities causes significant health issues among women and children [4].

It is essential to monitoring progress toward realizing SDG 6 for effective policy development and implementation. However, challenges remain in systematically assessing progress at national and local levels due to lack of standardized indicators and data collection methods [5]. Economic development is related to water and sanitation indicators. A rise in per capita GDP correlates with better sanitation outcomes. This reduces water-related illnesses [6]. Thus, there is a need for integrated approaches (that consider economic, social, and environmental factors).

The role of the private sector in achieving SDG 6 is increasingly recognized. Businesses can contribute to water management and sanitation solutions through innovation and investment [7]. However, it is crucial for governments to creating an enabling environment that encourages responsible business practices while also enacting legislation that holds companies accountable for their environmental and social impacts [7]. The collaboration between public and private sectors, along with civil society, is essential for addressing the multifaceted challenges associated with water and sanitation.

Climate change poses significant threats to the attainment of SDG 6, as it exacerbates water scarcity and affects the quality of water resources [8]. Extreme weather events, such as floods and droughts, can disrupt water supply systems and sanitation infrastructure, making it imperative to building resilience into water management practices [9]. Sustainable water supply systems must be designed to withstand climate-related disruptions, ensuring that communities have reliable access to clean water even in the face of environmental challenges [9].

2 The Importance of Water Sustainability in Global Development

Water sustainability is critical to global development. This is in the context of achieving Sustainable Development Goal 6 (SDG 6), which aims to ensure availability and sustainable management of water and sanitation for all. Water sustainability extends beyond access to clean water; it encompasses the efficient use of water resources, the protection of aquatic ecosystems, and the integration of water management with economic and social development. Water sustainability involves the rational utilization of water resources to meet the needs of the growing global population while ensuring the protection of ecosystems. According to the 2015 United Nations World Water Development Report, by 2025, two-thirds of the world's population could face water shortages, highlighting the critical need to adopt sustainable water management practices to prevent widespread crises and ensure equitable access to this vital resource [10].

One of the primary reasons for prioritizing water sustainability is its direct impact on public health. Access to clean water and adequate sanitation is crucial for preventing waterborne diseases, which disproportionately affect vulnerable populations, particularly in developing countries. The World Health Organization (WHO) estimates that unsafe drinking water and inadequate sanitation contribute to approximately 485,000 diarrheal deaths each year [11], as shown in Fig. 2.

Furthermore, studies have shown that improving water and sanitation access can significantly reduce the burden of childhood diseases, enhancing overall community

Fig. 2 Consequences of contaminated water and insufficient sanitation [11]

health and productivity [12]. Thus, sustainable water management practices are vital for safeguarding public health and promoting social equity.

Moreover, the sustainability of global water use has been under scrutiny, with studies indicating a troubling trend of increasing reliance on non-sustainable water resources. highlight that surface water over-abstraction has been rising, particularly since the late 1990s, despite the availability of larger water supplies [13]. This situation poses significant risks not only to water availability but also to the health of ecosystems that depend on these water sources. Effective management strategies must therefore address both the quantity and quality of water resources to ensure long-term sustainability.

Water pollution is another pressing issue that threatens water sustainability, particularly in developing countries where access to clean water is already limited and emphasizes that water pollution significantly impacts public health and economic development, making it imperative to adopting sustainable practices that mitigate pollution and protect water quality [14]. The critical link between water quality and economic growth is reinforced by evidence showing that clean water significantly impacts health, education, and social equity, driving sustainable development across multiple sectors [15]. Thus, addressing water pollution is essential for achieving broader sustainable development goals.

Furthermore, the integration of sustainable water management practices is crucial for food security and agricultural sustainability. As highlighted by lortyom et al. [16], the nexus between food and water security is complex and requires coordinated efforts to ensure that agricultural practices do not deplete water resources. Sustainable irrigation practices, as discussed by Glavan et al. [17], are vital for maintaining agricultural productivity while protecting water resources. This is particularly important in the face of climate change, which poses additional challenges to water availability and agricultural sustainability.

In addition, environmental sustainability is another critical dimension of water sustainability. The degradation of freshwater ecosystems due to pollution and over-extraction poses significant risks to biodiversity and ecosystem services. Sustainable water management practices, such as the use of microalgae for nutrient removal in wastewater treatment, can mitigate these impacts by enhancing water quality and promoting the recovery of essential nutrients [18]. Microalgae have demonstrated the ability to assimilate inorganic nutrients and heavy metals, thereby reducing eutrophication and improving the overall health of aquatic ecosystems [19]. This approach not only addresses pollution but also supports the restoration of natural habitats, which are essential for maintaining biodiversity.

Furthermore, the interconnections between water, energy, and food systems highlight the need for integrated approaches to sustainability. The concept of the water-energy-food nexus emphasizes that decisions made in one sector can significantly affect the others. For example, inefficient water use in agriculture can lead to water scarcity, which in turn affects energy production and food security [20]. Therefore, adopting a holistic perspective that considers the interdependencies among these sectors is crucial for achieving sustainable development goals.

3 Introduction to Microalgae and Their Ecological Significance

Microalgae are unicellular or multicellular photosynthetic organisms that thrive in aquatic environments, including freshwater, marine, and extreme habitats as shown in Fig. 3. They play a fundamental role in the ecological dynamics of these ecosystems, primarily as primary producers. Microalgae contribute significantly to global primary production, accounting for approximately 45% of the total, which surpasses the combined output of all terrestrial plants [21]. This immense productivity underpins aquatic food webs, supporting a diverse range of organisms from zooplankton to fish and ultimately sustaining higher trophic levels, including humans [22].

The ecological significance of microalgae extends beyond their role in primary production. They are crucial in regulating biogeochemical cycles, particularly the carbon and nutrient cycles. Through photosynthesis, microalgae fix carbon dioxide and release oxygen, thus contributing to the global oxygen supply and playing a vital role in mitigating climate change [23]. Furthermore, microalgae are instrumental in nutrient cycling, particularly nitrogen and phosphorus, which are essential for the growth of aquatic organisms. Their ability to assimilate these nutrients helps prevent eutrophication, a process that can lead to harmful algal blooms (HABs) and subsequent declines in water quality [24].

Microalgae also serve as bio-indicators of environmental health due to their sensitivity to changes in water quality and pollution levels. For instance, variations in microalgal community composition can reflect shifts in nutrient availability and the presence of contaminants, making them valuable tools for monitoring ecosystem health [25]. Additionally, their interactions with bacteria in aquatic environments are critical for nutrient cycling and energy flow, as they can influence bacterial community structure and function through the release of organic compounds [26].

Fig. 3 Microalgae: life's building blocks

The adaptability of microalgae to various environmental conditions further enhances their ecological significance. They can thrive in a range of habitats, from cold polar regions to hot springs, and can even tolerate extreme conditions such as high salinity or pollution [27]. This resilience not only contributes to biodiversity but also ensures the stability and functioning of aquatic ecosystems in the face of environmental changes.

4 Microalgae in Wastewater Treatment

4.1 Overview of Wastewater Challenges

Wastewater has long been considered a critical component of environmental and public health. It poses a significant challenge both worldwide and on the African continent [28]. Because of the breakdown of the pollutants and the lack of proper treatment, the presence of pollutants including organic matter, nutrients, infections, and heavy metals poses a serious threat to the aquatic ecosystem, the marine habitat, and public health [29]. Ineffective wastewater treatment can result from the use of antiquated plants and treatment techniques, which allows contaminants to escape into the environment and endangers public health. Conversely, the growing urbanization brought on by civilization and population growth presents another significant difficulty [30]. In an attempt to meet population demand, industrialization is accelerated while ignoring the negative environmental effects. In addition, wastewater treatment requires a lot of energy and is costly to start and keep up. Due to the use of cheap, impolite, and badly maintained infrastructure, this may result in wastewater overflow and leakage, which could increase the amount of hazardous waste released into the environment. Moreover, the agencies in charge of environmental regulation ought to guarantee complete adherence to the treatment plans put in place. They must require the use of professionals, advanced treatment infrastructure, renewable energy, and wastewater treatment before discharge into the environment. They should strictly be placed on ban or even closure of the establishment if the environmental regulatory rule is breached.

4.2 Microalgae as a Solution for Nutrient Removal (Nitrogen, Phosphorus)

Microalgae are microscopic organisms with a high potential of removing contaminants from wastewater [31]. Microalgae, single-celled and photosynthetic organisms, are always present in a moist environment and aquatic habitat. They support environmental sustainability because of their capacity to remove impurities from

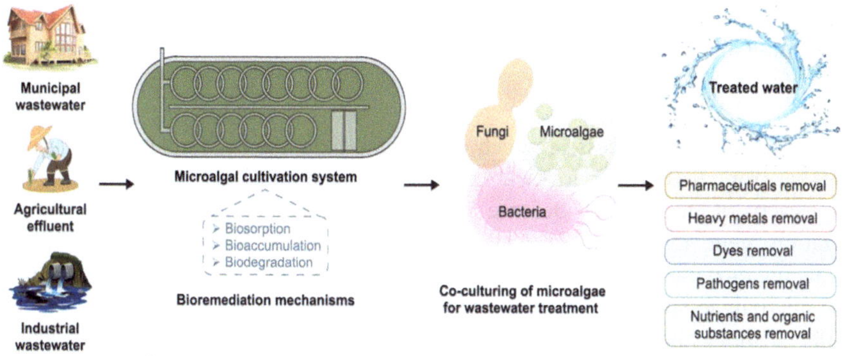

Fig. 4 Various categories of wastewater that can be treated with microalgae [36]

wastewater without releasing hazardous compounds [32]. The growth and reproduction of microalgae depend on the nutrients captured and other important factors. Microalgae are capable of removing nitrogen, phosphorus, and their compounds [33]. They absorb compounds such as nitrate, nitrite, and ammonium ions to generate proteins and other cellular components needed for their growth [34]. Phosphate helps in producing nucleic acid, adenosine triphosphate, and phospholipid [34]. Microalgae reduce the carbon footprint in the atmosphere through carbon sequestration. They remove contaminants from various categories of wastewater, including industrial, municipal, and agricultural wastewaters, before they are finally discharged into water bodies. The effectiveness of microalgae in removing nutrients suffers huge setbacks due to some important factors such as light intensity, temperature, nutrient ratio, pH, and concentration [35]. Sufficient sunlight accelerates the growth of the microalgae and the removal efficiency of the nutrients. A high temperature induces nutrient uptake and growth of the microalgae, while low to medium pH is efficient for the growth and removal of the nutrients by the microalgae, a high pH inhibits both the growth and nutrient-removal efficiency. The maximum removal of the nutrients and growth of the microalgae require a balance in the ratio of phosphorus to nitrogen because an unbalanced ratio of the nutrients may reduce the effectiveness of the process and the growth of the microalgae. Though microalgae can grow and remove nutrients at any concentration, high concentrations can lead to maximum removal of the nutrients and induce the growth of the microalgae (Fig. 4).

5 Microalgae in Industrial Effluent Treatment

In treating heavy metals and organic pollutants, microalgae have been known to be a promising removal of contaminants and pollutants from wastewater due to their environmental adaptation ability and tolerance range toward the various pollutants in the wastewater [36]. The mechanisms used by microalgae to remove heavy metals and pollutants include the following.

i. Biosorption

Biosorption is the process in which microalgae bind and accumulate heavy metals (HMs) from wastewater to their cell surface or intracellular compartment [37]. A variety of distinct self-defense mechanisms, including heavy metal immobilization, gene regulation, exclusion, and chelation, as well as antioxidants and reducing enzymes that lower heavy metals through redox processes, are employed by microalgae species to protect themselves against the toxicity of heavy metals [38]. Without altering their own activity, microalgae can generate organometallic compounds between heavy metals and proteins found in cells. Within vacuoles, the organometallic complexes are further divided to assist in controlling the concentration of heavy metal ions in the cytoplasm, thereby reducing their harmful effects [38].

ii. Biodegradation

Biodegradation is the term used to describe the breakdown of organic compounds caused by microorganism-produced enzymes. Co-metabolism or metabolic degradation are two possible ways. Organic compounds provide the only carbon and energy sources for metabolic breakdown. In contrast, nonspecific enzymes found in the environment are necessary for the breakdown of organic pollutants in co-metabolism because they catalyze the metabolism of other substrates. Organic micropollutants are typically found in wastewater in concentrations between ng/L and mg/L, which means that they might not be enough to support the growth of microalgae through metabolic breakdown. Consequently, co-metabolism might be the primary mechanism causing MP degradation. Microalgae are thought to be the most efficient means of removing organic MPs from the aqueous phase through biodegradation. Both external and intracellular biodegradation fall under this category. Simpler molecules are produced as a result of the catalytic degradation of complex parent chemicals by microalgal activity. Microalgae have a complex enzymatic system that allows for a variety of enzymatic interactions with the MPs during biodegradation, including ring cleavage, hydroxylation, carboxylation, oxidation, hydrogenation, and demethylation [39].

iii. Advantages over Conventional Treatment Methods

Microalgae wastewater treatments typically target the removal of ammonium, nitrates, and phosphorus, which are present in most wastewaters. Carbon is converted to CO_2 in conventional activated sludge treatment facilities, nitrogen (N) is removed and released into the atmosphere as N_2, and phosphorus (P) is often precipitated to prevent nutrient valorization [40].

The treatment of wastewater using microalgae is a more ecologically friendly option than conventional techniques. It obtains its energy from natural processes rather than chemicals. By doing this, the creation of sludge is greatly reduced, and the microalgal biomass that results can be used to make biofuel, fertilizer, and pharmaceuticals, among other things. This method encourages resource recovery and provides a solution to sludge disposal. Microalgae consume nutrients, which lowers the cost of growth maintenance. Additionally, the oxygen they produce during wastewater treatment provides an environment that is favorable for other microorganisms, lowering the need for aeration and its associated expenses [41]. The following are the typical cost savings attained when employing microalgae as a bioremediation agent for wastewaters:

a. Operation costs are decreased because microalgal growth is facilitated by the nutrients in wastewaters.
b. Microalgae just need sunlight and CO_2 from the atmosphere or flue gas to perform photosynthesis.
c. Microalgae produce oxygen rather than consume it. Aeration costs are decreased because bacteria use this oxygen to further degrade the material.
d. Microalgal biomass is suited for use as fertilizers and reduce the fertilizer cost of agro-industries.

6 Innovations in Microalgae Technology for Water Purification

i. Emerging Technologies and Methodologies

It is necessary to create innovative technologies and ecologically friendly ways in order to completely understand and address the water quality challenge. The improved treatment efficiency of nanotechnology, which under some circumstances enables cost-effectiveness, and its acceptance as a therapeutic approach make it a potentially valuable technology. Delivering high-performance, affordable wastewater treatment while reducing reliance on existing infrastructure is made possible by nanotechnology. It guarantees more economical and sustainable resource consumption [42]. Electrochemical systems employ applied electrical currents to drive separations in bulk electrolytes, electrochemically trap impurities in electric double layers (EDLs), or intercalate impurities in solid electrodes in order to eliminate them from the feed. Electrokinetics and electrosorption are the two primary techniques used in an electrochemical process to eliminate impurities. Electrokinetic process is usually a continuous process that depends on efficient mass transfer to remove pollutants. It involves the movement of charged or uncharged but dielectric species in an electrolyte in response to an applied electric field. Electrosorption process is a cyclical process that includes all events where an applied electric field facilitates the binding of pollutants. Electrosorption depends on favorable reaction kinetics and thermodynamics in addition to efficient transport [43].

ii. Integration with Existing Water Treatment Systems

The European Commission has identified anaerobic digestion of organic wastes as one of the most energy-efficient solutions for harnessing biological wastes' energy potential. It is theoretically and financially possible to combine the production of microalgae with wastewater treatment to produce biofuel by anaerobic digestion or co-digestion with other substrates. Additionally, using microalgae to produce energy improves process efficiency, lowers WWTP costs, and promotes system viability. Microalgae that are co-digested with a carbon-rich substrate, like waste activated sludge (WAS), produce a combination with an acceptable C/N ratio and are not inhibited by ammonia (since microalgal biomass has a high protein content) or volatile fatty acids (VFAs). Co-digestion also improves alkalinity and process stability, balances macro- and micronutrients, increases the load of organic matter that can be broken down by bioprocessing, boosts methane productivity, and improves hydrolysis efficiency while lowering the danger of inhibition [44].

7 Biomass Production from Wastewater Treatment

i. Utilizing Treated Biomass for Biofuel, Fertilizers, and Other Products

Microalgae can develop in a variety of trophic conditions. Microalgae, for instance, can develop photoautotrophically to absorbing CO_2 from the atmosphere, which reduce the amount of CO_2 and produce valuable products, heterotrophically, which results in the extraction of nutrients from the wastewater and the creation of biofuel, or mixotrophically, which simultaneously captures CO_2, treats wastewater, and produces valuable goods [45].

ii. Economic and Environmental Benefits

Microscopic aquatic creatures known as microalgae have several advantages for the environment and economy. They are a viable resource for sustainable development because of their quick growth, capacity to flourish in a variety of environments, and ability to produce useful products. The application of microalgae in biofuel, biofertilizer, cosmetic, medicine and health, agriculture, waste treatment, and food and additives shows the economic importance of the microscopic organism. It increases the sources of revenue for industry, nations, and scientific research where it is being utilized. They relieve pressure on water and land resources because they are a resource that is renewable and adaptable to different climates [46]. Microalgae can also help slow down the effects of climate change because they can produce biofuels and absorb carbon dioxide from the atmosphere. Additionally, they can safeguard aquatic habitats, increase biodiversity, and remove contaminants from wastewater to improve the quality of the water.

8 Enhancing Resource Recover

i. Role of Microalgae in Recovering Valuable Resources from Wastewater

Microalgae's remarkable efficiency in removing pollutants from wastewater from domestic, industrial, and agriculture has led to numerous studies demonstrating its viability as an advanced wastewater treatment technology [47]. Microalgae can contribute to pollution control by absorbing nutrients such as nitrogen and phosphorus from wastewater. They also help in carbon sequestration by capturing atmospheric carbon dioxide due to their photosynthetic ability. Microalgae helps to recover heavy metal pollutants, organic matters, and pharmaceutical contaminants, which can have adverse effects on aquatic ecosystem. This alone improves the quality of water. The biomass produced during the growth phase of microalgae can be harvested and used in the production of important and valuable products including biofuel, biofertilizer, and pharmaceutical and dietary products. The adoption of microalgae in wastewater treatment is cost-effective and environmentally friendly.

ii. Contribution to a Circular Economy in Water Management

A circular economy designs systems and products that are reusable, recyclable, or repairable in order to reduce waste and increase resource efficiency, and it differs from the conventional linear economy, which extracts, uses, and discards resources. Preserving stocks' long-term worth has always been the aim of the circular economy, regardless of whether they are made of cash, people, culture, natural resources, or manufactured items [48]. Due to the fact that wastewater is full of valuable materials that may find secondary uses that would keep them in the loop as long as possible while lowering the quantity of waste generated, CE has become especially significant in the urban water cycle, especially in the wastewater treatment industry [49].

Restoring water resources and closing the gap between supply and demand can be achieved by treating wastewater to the proper quality. In locations where water is scarce, treated wastewater offers an alternate source of water for irrigation, industrial applications, and potable supply. Subject to the necessary treatment, water can be recycled for a variety of purposes including agriculture (such as food crops, hydroponics, aquaculture, and greenhouses), industry (such as processing water, cooling water, and recirculating cooling towers), urban uses (such as irrigation of public parks, sporting facilities, private gardens, and roadsides), recreation (such as golf course irrigation), environmental (such as aquifer recharge; wetlands; marshes), and potable (such as aquifer recharge for drinking water use; augmentation of surface drinking water supplies) [50].

9 Microalgae and Water Security

Owing to the metabolic adaptability of microalgae, which allows them to execute photoautotrophic, mixotrophic, or heterotrophic metabolism [51], microalgae are potential biological systems for treating a wide range of wastewater sources as highlighted above, specifically, in the framework of a circular and bio-based economy, and the development of biorefinery concepts [52]. Given the growing worldwide need for freshwater, it is especially vital to concentrate on desalination in the context of microalgae. With freshwater resources becoming scarcer, desalination processes are vital for providing potable water from seawater and brackish water. Microalgae provide a sustainable and eco-friendly alternative to desalination procedures, potentially lowering energy consumption and environmental impact compared to existing technologies such as reverse osmosis and thermal desalination. Although microalgae have the potential to eliminate many pollutants, the pressing need to tackle water scarcity makes desalination a crucial field of research. The use of microalgae in desalination involves salt biosequestration and the generation of bio-based components for use in desalination membranes. By focusing on desalination, the study addresses the global demand for new and sustainable water purification options, making it both topical and impactful.

10 Microalgae in Desalination Processes

10.1 Potential Role of Microalgae in Reducing the Environmental Impact of Desalination

In order to meet the urgent requirement for sustainable desalination processes, which are essential for producing drinkable water from saltwater and controlling highly saline wastewater (HSWW) from diverse sectors, microalgae-based biodesalination offers a viable alternative. Traditional desalination processes such as reverse osmosis (RO), multi-effect distillation, and flash distillation, while effective in providing drinking water, have substantial environmental and economic downsides. These methods are energy-intensive, emit significant levels of carbon dioxide (CO_2), generate enormous amounts of brine, and require chemicals for membrane cleaning, which can impact natural resources and ecosystems [53]. Furthermore, the handling of by-products such as brine and sludge, as well as the acquisition of corrosion-resistant materials for waste disposal, increases the cost of these systems. In this context, the development of biodesalination techniques based on microalgae presents a compelling option that could reduce the environmental impact of traditional desalination. Microalgae, particularly halophilic strains, have demonstrated significant promise for lowering the salt content of saline water by natural biological processes. These processes, which include bioadsorption and bioaccumulation, enable microalgae to absorb salts from the water as nutrients for development. Biosorption, in example, is an energy-independent process that has been identified as the major mechanism in biodesalination, making it a low-cost and ecologically benign approach for creating freshwater from seawater or brackish water [54]. One of the most promising microalgae strains for biodesalination is ***Picochlorum***, which is noted for its fast growth rate, high salt tolerance, and ability to survive in high CO_2 settings. These qualities make ***Picochlorum*** an excellent choice for large-scale biodesalination applications. The ability of microalgae to adapt to high saline conditions and efficiently remove salts from water not only provides a sustainable form of desalination, but it also tackles the issues connected with HSWW management. HSWW is generated by industries such as food processing, metallurgy, coal, chemical, and leather production, and if not treated, it can have serious effects on human health and the environment. High salt intake from contaminated water sources can raise blood pressure, increase the risk of stroke, and contribute to kidney illness, coronary heart disease, and osteoporosis [55].

The use of microalgae in desalination operations can also minimize dependency on energy-intensive technology, lowering overall carbon footprint. The environmental impact is further reduced by utilizing microalgae's natural ability to sequester CO_2 during the desalination process. Furthermore, microalgae-based biodesalination does not generate toxic by-products such as brine, which is a major concern in traditional desalination processes. This not only makes the process more sustainable, but it also mitigates the difficulties connected with brine disposal and its harmful impacts on marine ecosystems [56, 57]. Furthermore, the use of microalgae in desalination is consistent with the global trend toward developing green technologies that are both effective and environmentally friendly. As law in many countries tightens the treatment of industrial wastewater, the use of biodesalination methods could assist enterprises in complying with these regulations while minimizing their environmental effect. The capacity of microalgae to remove both cationic and anionic contaminants from liquid environments increases their potential for wastewater treatment and desalination applications [58]. Recent advances in bioprospecting have found unique microalgae strains with extraordinary features, including rapid doubling periods and severe halophilicity, which have the potential to revolutionize biodesalination [59]. The continuing investigation and development of microalgae strains such as ***Picochlorum*** may pave the way for more efficient and sustainable desalination procedures, ultimately benefiting our environment and human populations.

10.2 Innovation in Algae-Assisted Desalination Technologies

Microalgae-assisted desalination technologies are emerging as a sustainable and novel approach to addressing water scarcity by leveraging microalgae's intrinsic properties. These systems combine biological processes with desalination methods to improve efficiency, reduce environmental impact, and lower energy costs. Pre-treatment innovations such as algal bioreactors help to remove pollutants while minimizing membrane fouling and maintenance expenses. Algae can absorb salts and produce biofuels, making them more energy-efficient than other desalination techniques. Furthermore, algae can process desalination by-products like brine, transforming waste into valuable biomass for biofuels or fertilizers. This approach increases desalination efficiency while reducing environmental impact and operational costs. These technologies use biological agents including bacteria, enzymes, and cells to trigger biochemical reactions in different materials. Among the several bio-based approaches, the conventional activated sludge (CAS) process is frequently utilized. CAS is extremely effective in removing organic pollutants and nutrients from wastewater; however, it has some limitations, including high energy consumption and foam generation, which creates operational issues and increases management costs [60]. Furthermore, the formation of waste-activated sludge (WAS) increases treatment costs, accounting for up to 60% of total wastewater treatment facility operational costs [61]. In light of these issues, integrating algae with wastewater treatment technology has become a top priority. Microalgae-based systems, which can remove fertilizers such as nitrogen and phosphorus while creating oxygen via photosynthesis, offer a promising alternative to CAS. These systems have substantially higher energy efficiency since they rely on photosynthetic oxygen generation rather than mechanical aeration, and their potential to reduce CO_2 emissions makes them a more sustainable solution [62]. However, biomass separation in microalgae systems remains a significant difficulty since separating algal biomass from treated water is challenging [36].

The answer to these problems has resulted in the invention of Algae-Based Membrane Bioreactors (AMBRs). This novel technique combines the benefits of algae photosynthesis and microbial decomposition with membrane filtration to improve wastewater treatment. AMBRs use microalgae and bacteria in a bioreactor, with membranes acting as a barrier to separate treated water from biomass. This structure not only assures excellent pollutant removal, but it also helps to retain high concentrations of biomass, addressing the biomass separation difficulty that standard microalgae systems confront [63]. AMBR technique has gained popularity in tropical and subtropical climates, where sunshine availability increases algae photosynthesis, lowering the energy costs associated with mechanical aeration [63]. The combination of algae and bacteria in these reactors improves nutrient removal, notably nitrogen and phosphorus, which are major contributors to water pollution and eutrophication in natural water bodies. Furthermore, algae in these systems can absorb CO_2 produced by bacterial activities, resulting in large reductions in greenhouse gas emissions [64]. This technique, called **photosynthetic oxygenation**, is a significant advantage of AMBRs over typical wastewater treatment methods. Membrane Bioreactors (MBRs) are becoming increasingly popular in wastewater treatment due to their superior performance in organic load management, reduced sludge production, and small operational footprints. The use of membranes in bioreactors improves treated water quality by allowing for efficient solid–liquid separation and biomass retention inside the system [65]. MBRs are particularly successful in breaking down organic pollutants, but their nutrient-removal capacity, notably for nitrogen and phosphorus, is sometimes limited [66]. This constraint has motivated the development of microalgae-based membrane bioreactors (AMBRs), which combine the benefits of MBR technology with the nutrient-removal abilities of microalgae. Recent research has demonstrated that AMBRs can reach considerable pollutant removal efficiency, making them a potential solution for municipal and industrial wastewater treatment. The introduction of microalgae into MBR systems has various benefits, including better CO_2 fixation, nutrient removal, and resource recovery. Microalgae can absorb nitrogen and phosphorus from wastewater while also producing useful biomass for biofuel generation or other industrial applications [67].

Furthermore, AMBR systems have been shown to remove heavy metals from industrial wastewater, increasing its usefulness in environmental management [68]. Recent advances in AMBR technology have centered on optimizing membrane materials, reactor layouts, and operational parameters in order to increase treatment efficiency and lower operational costs. Membrane fouling is one of the most serious difficulties in AMBR systems, reducing membrane lifespan and increasing maintenance costs (L. [69]). To address this issue, researchers have created novel membrane materials, such as polyvinylidene fluoride (PVDF), that are very resistant to fouling and chemical degradation [70]. Furthermore, developments in membrane module design, including the development of flat-sheet and hollow-fiber designs, have boosted the performance of AMBR systems by enhancing mass transfer and reducing energy consumption. Another area of study has concentrated on enhancing CO_2 mass transfer in AMBR systems, which is critical for maximizing algal growth and CO_2 fixation as shown in Fig. 5 (L. [71]). The poor solubility of CO_2 in water makes it difficult for microalgae to properly use the available CO_2. Innovative reactor designs, such as hollow fiber membrane photobioreactors (HFMPBs), have been created to improve gas–liquid interfacial contact, thus increasing CO_2 utilization and algal production [72]. The combination of forward osmosis (FO) and AMBR technology is another promising approach to enhancing wastewater treatment and resource recovery. FO uses a semi-permeable membrane to transport water over a concentration gradient, effectively concentrating microalgae while lowering energy costs associated with traditional dewatering techniques. This technology can be used in conjunction with AMBRs to enhance water recovery, nutrient removal, and algal biomass harvesting.

Other novel approaches include the Microbial Desalination Cell (MDC) system. In an MDC system as shown in Fig. 6a and b [73], electroactive bacteria oxidize organic materials in wastewater, producing electrons that pass to the anode and generate an electrical current. The gradient between the anode and cathode chambers drives this current, which powers the desalination process. Unlike other microbial electrochemical technologies (METs), MDC has an additional desalination compartment between the anodic and cathodic compartments, which improves its capacity to desalinate water [74]. Despite its intriguing potential, MDC confronts hurdles, particularly the slow kinetics of the oxygen reduction reaction at the cathode, which reduces the efficiency of freshwater generation and salt removal. Various electron acceptors, including oxygen, ferricyanide, and permanganate, have been explored to improve system efficiency; however, their depletion affects sustainability [75].

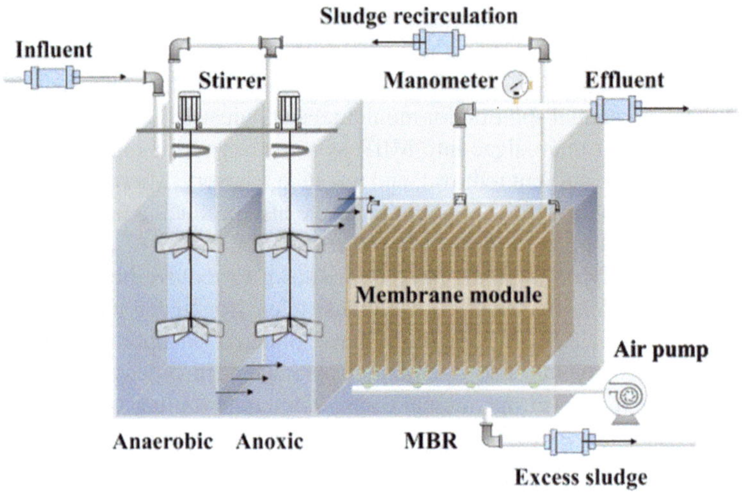

Fig. 5 The schematic of the pilot-scale limited-aeration MBR [71]

Recent research has focused on the utilization of biocathodes, in which microorganisms influence reduction processes, providing a cost-effective and natural buffering option for salt removal and wastewater treatment [76]. However, algae-assisted MDC systems, which were introduced in 2013, continue to suffer with poor current generation and desalinization rates. Research efforts, such as the research of Chlorella vulgaris in MDCs, are continuing to improve the system's electrochemical behavior, current efficiency, and overall desalination performance, paving the way for further optimization of algae-assisted MDCs [77].

11 Microalgae's Role in Enhancing Water Quality in Natural Water Bodies

The inclusion of microalgae in natural water bodies such as lakes, rivers, and seas is of paramount importance since they effectively mitigate pollution and uphold ecological equilibrium. These microbes enhance water quality by absorbing nutrients like nitrogen and phosphorus, which are frequently responsible for eutrophication and hazardous algal blooms. [78] describe phycoremediation, which uses microalgae to remove pollutants from water systems, as a natural and sustainable technique of purification. Microalgae protect water bodies from oxygen deprivation by preventing the growth of toxic algae and eliminating excess nutrients. Unlike traditional wastewater treatment methods, microalgae-based systems are both cost-effective and environmentally beneficial, as they produce no toxic by-products or secondary contamination [79]. Furthermore, microalgae do not compete with agricultural land, making them an appealing approach for enhancing water quality while

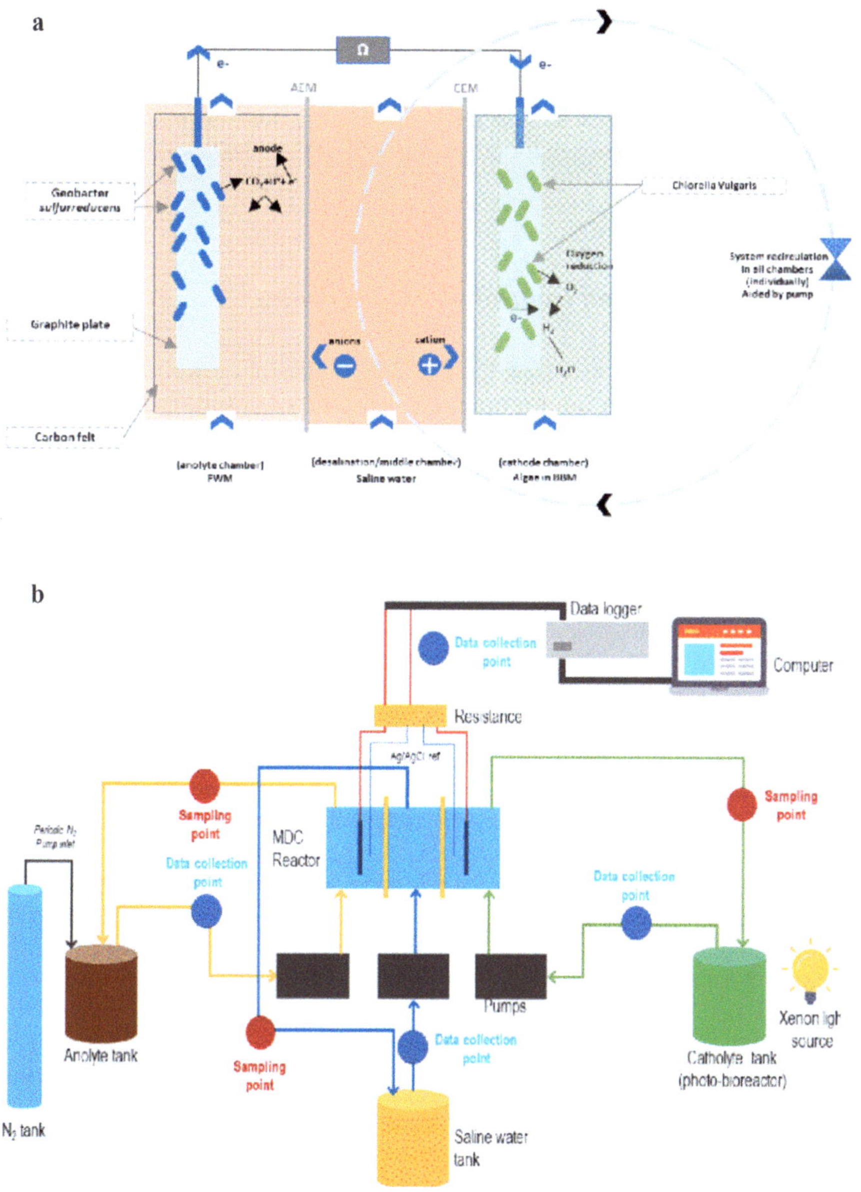

Fig. 6 (a) Details of reactions and compartment of MDC cell [73] **b** Diagram showing the experimental setup and flow pattern [73]

maintaining food production [80]. Furthermore, incorporating microalgae into water management systems enables the simultaneous production of biomass, which may be utilized to generate biofuel, bringing an energy-efficient dimension to water treatment efforts [81]. By focusing on nutrient removal and ecosystem restoration, microalgae provide a preventive approach for protecting the health and quality of natural water systems.

11.1 Prevention and Control of Algae Blooms

The prevention and control of algae blooms, a phenomena characterized by excessive algae growth in water bodies, has become a major environmental concern due to its influence on ecosystems, water quality, and human health. Over the years, researchers have devised a variety of strategies, and multiple review articles have been published that focus on suppressing high-density blooms by inactivating algae. [82] emphasized the function and mechanisms of microbial species in moderating algae blooms, which provided insight into biological management approaches. [83] examined the effectiveness, processes, and influencing factors of several control methods, emphasizing options for managing algal blooms based on their initial density. This review also highlighted innovative techniques, such as locally enhanced electric field treatment (LEEFT) and magnetic nanoparticles, which can trap and remove algae at various density levels [83]. Algal bloom control strategies can be classified as physical, chemical, physicochemical, biological, and ecological. Physical methods such as mechanical harvesting, aeration, air floatation, and membrane separation provide quick, short-term solutions, but they are frequently limited by cost, maintenance, and efficiency. Mechanical harvesting, for example, swiftly removes vast amounts of algal biomass but is restricted by the size and cost of the apparatus involved [84]. Similarly, aeration raises dissolved oxygen levels, which inhibits algal development by restricting photosynthesis, whereas air floatation technology effectively removes algae by forming flocs that float to the water's top. However, these systems are frequently limited by high operational costs and the requirement for ongoing maintenance [85]. Membrane separation has emerged as another viable approach for algae control, despite obstacles such as fouling by algae particles and extracellular organic materials. Innovations such as vibrating membrane separation based on frequency control have increased efficiency, but the approach remains prohibitively expensive for long-term use as shown in Fig. 7 (S. [86]). Ultrasound is another novel physical method that has the potential to inactivate algae by damaging cell membrane structures, inhibiting photosynthesis, and producing reactive oxygen species. Although ultrasound can effectively suppress algal development, issues about energy consumption, water quality effects, and the release of toxins from algae cells warrant more investigation [87].

Algal blooms are commonly controlled chemically, with algaecides such as copper sulfate, potassium ferrate, and hydrogen peroxide. Copper-based algaecides, for example, have been shown to effectively prevent algal growth, but when applied in

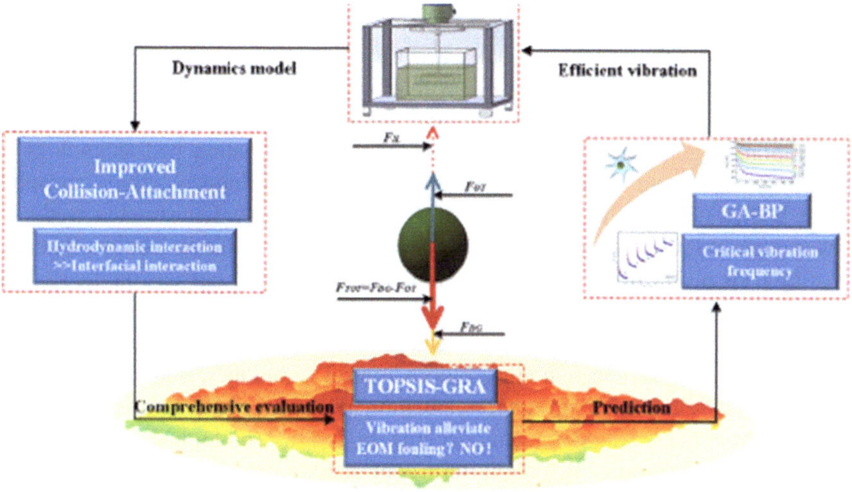

Fig. 7 The vibrating membrane separation method for algal bloom control [86]

high quantities, they can produce disinfection by-products and pose environmental risks [88]. Potassium ferrate has been shown to damage cyanobacteria cell structures while reducing the possibility of toxic consequences. However, using chemical treatments carries dangers such as the quick release of algal toxins upon cell demise, as well as potential long-term consequences on water quality and human health [89]. These hazards highlight the significance of carefully selecting chemical agents, with a preference for those that leave few residues in the environment [90]. Physicochemical strategies, which mix physical and chemical approaches, have proven effective in algae control. Coagulation-air floatation and flocculation-sedimentation are two such techniques. Coagulation-air floatation involves coagulating algae cells using surfactants such as aluminum sulfate or ferric chloride before removing them from the water using air floatation. This approach has shown excellent efficiency, with clearance rates of up to 99% under ideal conditions [91]. However, practical issues such as the generation of large dissolved bubbles can impair the efficacy of air floatation [92]. Similarly, the flocculation-sedimentation process, which consolidates microscopic particles into larger ones for easy removal, is a possibility. However, selecting flocculants and adjusting system pH are crucial to guaranteeing efficacy [93].

Biological and ecological strategies provide a more ecologically friendly solution to algal bloom control. These approaches use bacteria, aquatic creatures, or plants to control algal development. Microorganisms such as bacteria and fungi have demonstrated long-term inhibitory effects on algal cells, with certain strains, such as the chitinase-producing bacterium LY03, obtaining a 70% algicidal rate after 24 h [94]. Aquatic species such as tilapia have also been shown to consume huge amounts of algae, providing a natural regulating mechanism [95]. Furthermore, aquatic plants can restrict algal growth by competing for nutrients, shading, or emitting allelopathic substances that impair algae cell structure and function. For example, Fig. 8

shows the principles and approaches of allelopathic plants to inhibit bloom-forming microalgae [96] while Arundo donax extract has been shown to be effective in inactivating ***Microcystis aeruginosa*** cells, with efficacy dependent on the initial density of the algae [97]. The goal of ecological approaches such as aquatic vegetation restoration, engineered wetland technology, and plant floating bed technology is to restore ecosystem equilibrium. Planting nutrient-absorbing plants in eutrophic water bodies reduces the availability of nutrients for algae. Constructed wetlands lower nutrient levels through plant adsorption or microbial mineralization, whereas floating beds purify water bodies by fixing plants in floating carriers on the surface [98]. Although these procedures take longer to take impact than physical or chemical approaches, they are regarded more sustainable and environmentally beneficial in the long run. Despite advances in algae bloom control techniques, bottlenecks persist. The high costs, environmental dangers, and limited long-term efficacy of physical and chemical treatments are significant concerns. In contrast, biological and ecological approaches, while promising for long-term control, require additional refining and are frequently too sluggish to be practical in large-scale outbreaks. Integrating several approaches, such as combining ultrasonography with biological control measures or employing ecological restoration with chemical treatments, may provide a more comprehensive answer to this complicated issue [99].

Future research should concentrate on establishing more efficient, cost-effective, and environmentally friendly control approaches. New technologies, such as LEEFT and magnetic nanoparticles, offer promise, but their scalability and long-term impact require further exploration. Furthermore, as ultrasonic control technology progresses, attention must be devoted to optimizing frequency, power, and exposure duration to reduce energy usage while increasing effectiveness. Researchers want to develop more sustainable and economical techniques for controlling algal blooms in the future by improving existing technology and investigating new approaches. Therefore, controlling algae blooms necessitates a complex strategy that blends short-term solutions with long-term ecological balance. Mechanical harvesting and membrane separation, chemical algaecides, and innovative technologies such as ultrasound and

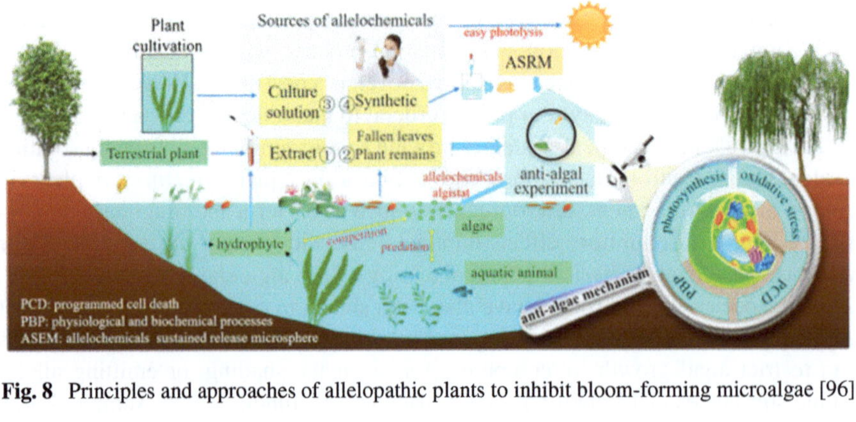

Fig. 8 Principles and approaches of allelopathic plants to inhibit bloom-forming microalgae [96]

LEEFT all provide quick relief, but they come with costs, maintenance, and environmental dangers. Biological and ecological methods, on the other hand, provide environmentally sustainable alternatives, although more research is needed to improve their practicality and efficiency. In the next years, effective algae bloom management will require the continued integration of current strategies, as well as the development of new technology.

11.2 Microalgae in Bioremediation of Polluted Water Bodies

Microalgae play an important role in improving the quality of natural water bodies, especially as global pollution levels rise and the demand for clean water grows. Water pollution has increased since the Industrial Revolution as a result of rapid industrialization and population growth. The agriculture industry, for example, consumes large amounts of pesticides and generates a lot of organic waste, which contributes to water contamination [100]. Pollutants such as persistent organic pollutants (POPs) and heavy metals have raised serious concerns due to their toxicity and ability to bioaccumulate in aquatic environments [101]. These contaminants, coupled with emerging contaminants (ECs), are severe concerns to water quality and public health [102]. Several water treatment procedures, including coagulation, filtration, and oxidation [103], have been developed, but large-scale deployment remains a barrier. In response, microalgae have emerged as a possible alternative due to their distinct biological properties and ability to remove contaminants from wastewater [104]. Microalgae are a diverse species of unicellular photosynthetic organisms that may grow in a variety of wastewater settings. Their cell walls contain complex functional groups capable of binding contaminants via biosorption, a mechanism that is independent of algal metabolism and works on both living and dead biomass [105]. Living microalgal cells can absorb and biodegrade specific substances in addition to biosorption, hence improving their total pollution removal capabilities [106]. The ability of microalgae to remove a wide range of contaminants, including nutrients, heavy metals, and organic pollutants, makes them an appealing choice for improving water quality. The ability of microalgae to use organic and inorganic carbon, nitrogen, and phosphorus is especially useful for wastewater treatment. According to [107], microalgae such as ***Chlorella vulgaris*** may remove up to 85% of total phosphorus (TP) and 89% of total nitrogen (TN) from wastewater. This nutrient-removal capability not only prevents eutrophication but also lowers the water's chemical oxygen demand (COD), increasing overall quality [108].

Microalgae, including ***Chlorella sp.***, effectively remove heavy metals from wastewater by biosorption [109]. These features make microalgae an effective tool for tackling organic and inorganic contaminants in aquatic systems. Because of their versatility and effectiveness, many microalgae species are widely used in wastewater treatment. ***Chlorella*** microalgae have been extensively researched for pollution removal because to their rapid growth rate and strong biosorption capacity [110]. ***Chlorella*** species can remove metal ions, including Al^{3+}, Ca^{2+}, and

Mg^{2+}, demonstrating their adaptability in treating various pollutants [111]. Similarly, ***Scenedesmus*** species, another often utilized microalgae, have demonstrated high efficacy in removing nitrogen, phosphorus, and suspended particles from industrial wastewater [112]. This versatility enables microalgae to treat a wide spectrum of wastewaters, from municipal to industrial sources, resulting in cleaner natural water bodies. Nutrient concentration, pH, and light availability all have an impact on microalgae's performance in wastewater treatment. Nitrogen and phosphorus levels in wastewater have a substantial impact on microalgal growth and pollutant removal effectiveness. Nitrogen intake is particularly boosted by the presence of phosphorus, and microalgae can participate in luxury phosphorus uptake, storing excess phosphorus for later use. pH levels also influence nutrient availability and the solubility of substances such as ammonium and phosphate. For example, high pH levels might cause the production of calcium phosphate, making phosphorus unavailable for microalgae uptake. Furthermore, the color and opacity of wastewater affect light absorption, which is required for microalgal photosynthesis. High-absorbance wastewaters, such as those from cattle or pulp and paper production, can inhibit microalgal growth and nutrient removal. As a result, optimizing these parameters is critical to increasing the effectiveness of microalgal-based water treatment systems.

Beyond pollutant removal, microalgae have the ability to produce biomass, which can then be turned into biofuels, fertilizers, and other important items. For example, ***Botryococcus braunii*** is a slow-growing algae capable of removing nitrogen and phosphate from wastewater while also creating hydrocarbons for biofuel production [113]. This dual functioning of microalgae contributes to the concept of a circular economy, in which waste is converted into resources, hence increasing the sustainability of water treatment procedures. Commercial uses of microalgal wastewater treatment are emerging, with businesses such as Algae Systems LLC and Algal Enterprises treating wastewater and producing biogas using floating photobioreactors (PBRs) and anaerobic digesters [114]. Other firms, such as Microbio Engineering and HydroMentia, use open raceway ponds and immobilized systems to remove nutrients and generate biomass [115]. While these systems show promise, the large-scale deployment of microalgae in wastewater treatment is still in its early stages, with more study needed to optimize growing conditions, nutrient-removal efficiency, and cost-effectiveness.

12 Challenges and Limitations

12.1 Technical and Operational Challenges

Microalgae is beneficial in high-rate algal ponds (HRAP) for wastewater treatment. It lowers greenhouse gas (GHG) emissions and operating costs. However, large-scale deployment of microalgae systems confronts a number of technological and operational hurdles that must be overcome before being used. These problems include land use concerns, environmental implications, and the technological difficulty of maintaining ideal growing conditions for microalgae. Large installation space is an obstacle of HRAP systems. These ponds, which use the growth of microalgae to treat wastewater, require broad open regions to function effectively [116]. The change in land use (LUC) to support HRAP systems can have negative environmental consequences. Converting woods, grasslands, or croplands to HRAP installations causes soil carbon loss, carbon removal below the original vegetation, and changes in surface albedo. These contribute to increased GHG emissions. Carbon sequestration potential of these natural ecosystems will be lost, resulting in a "carbon penalty." [117] found that converting grassland or cropland to microalgae ponds causes penalties of 4–8 g CO_2 eq/MJ, and converts forest land to penalties reaching 40 g CO_2 eq/MJ. Displacing ecosystems disrupts carbon sequestration from natural systems, which would otherwise capture atmospheric CO_2 [118].

Non-arable or marginal lands could be ideal for HRAP installations; the closeness of the area to wastewater generators complicates this method. The majority of wastewater is generated in metropolitan areas with high land costs. Transporting wastewater to remote regions where property may be cheaper or less valued for other purposes is not a practical solution due to significant transportation infrastructure costs [119]. This provides a logistical and economic challenge that must be overcome before large-scale HRAP systems can be considered as a viable replacement for conventional wastewater treatment systems.

Furthermore, the heterogeneity in environmental affects generated by LUC means that HRAP installations may have quite diverse effects depending on location. [117] found that, despite equal irradiance and temperature, LUC consequences can differ greatly depending on local environmental circumstances. Another significant operational problem is maintaining the appropriate environmental conditions required for microalgae to thrive. Temperature, light availability, pH, and nutrient concentration are all important factors in the success of HRAP systems. Microalgae operate at temperatures between 20 and 30 °C, outside of this range, growth rates and nutrient-removal efficiency can suffer. In cooler temperatures, metabolic rates slow, limiting the system's ability to effectively process wastewater. Overheating, on the other hand, can be troublesome, especially in humid areas where evaporation is impeded, resulting in lower photosynthetic activity and oxidative stress [120]. Implementing temperature control measures, such as providing cooling water or installing refrigerated air conditioning units, would dramatically raise operational expenses. Similarly, lighting intensity must be properly controlled. While increasing light can promote microalgal growth, excessive irradiance can cause photoinhibition, lowering photosynthetic rates and biomass yield. The appropriate mixing of the culture media is critical because it allows microalgae to be exposed to light intermittently, reducing photoinhibition and promoting even growth ([121]). However, excessive mixing can cause shear stress and cell injury, particularly in some species, complicating system performance [122]. In terms of nutrient concentrations, the carbon:nitrogen:phosphorus (C:N:P) ratio in wastewater is critical for microalgae-based systems. A shortage of critical nutrients, or their inadequate bioavailability, can reduce pollution removal effectiveness and biomass production. Furthermore, some wastewater chemicals, such as ammonia, can be hazardous to microalgae, especially at higher pH values. As the pH rises during photosynthesis, the concentration of free ammonia rises, inhibiting microalgal growth. To overcome this, strategies such as pH control, dilution, or low loading rates are required; nevertheless, these methods complicate system operation [123].

Hence, the economic and environmental implications of microalgae-based wastewater treatment must be thoroughly assessed prior to large-scale adoption. While pilot-scale HRAP systems have yielded encouraging results, a more thorough study of the costs, GHG emissions, and LUC implications is required. To fully analyze the viability of these systems, life-cycle studies that account for LUC emissions as well as long-term sustainability are required. Although HRAP systems provide an environmentally benign alternative to traditional wastewater treatment technologies, careful evaluation of location-specific characteristics, economic feasibility, and environmental implications is critical to their performance [80, 121]. While microalgae-based HRAP systems have significant potential for lowering GHG emissions and operational costs in wastewater treatment, they confront a number of technical and operational obstacles, including land use, environmental implications, and maintaining ideal growth conditions. Addressing these problems is critical for widespread implementation of this technology, which has the potential to dramatically improve sustainable wastewater management.

12.1.1 Scaling up Microalgae Systems for Widespread Use

When it comes to using microalgae in wastewater treatment, there are various obstacles and constraints to consider. A key concern is the possibility of harvested biomass being contaminated with dangerous bacteria, which might pose serious dangers if the biomass is repurposed, such as aquaculture feed [124]. The existence of both bacterial and microalgal biomass raises questions regarding the safety of using such goods in the absence of strict bacterial control. Furthermore, current harvesting technologies, such as floatation and fungi-assisted techniques, are neither commercially viable nor completely established for industrial application [125]. Floatation can introduce undesirable solid wastes into harvested biomass, lowering its safety, while the expenses of producing fungal pellets in artificial conditions are prohibitively expensive for commercial usage. When it comes to using microalgae for wastewater treatment, there are various obstacles and constraints to consider. Another significant constraint is the absence of detailed economic assessments and life-cycle analyses (LCA) for these systems, limiting the capacity to estimate their true environmental and economic implications. Without resolving these economic, environmental, and safety concerns, the large-scale application of microalgae systems is hampered, despite its potential benefits for wastewater treatment and aquaculture.

Also, microalgae can help reduce pollution from aquaculture by enhancing carbon dioxide fixation and wastewater clean-up. Their use can also assist reduce the developing issue of antibiotic resistance in aquaculture, resulting in a healthier and more sustainable ecology. Economically, microalgae's ability to collect nutrients from wastewater and convert them into biomass for aquaculture feed has the potential to greatly cut operational costs, as feed is a key expense in aquaculture. This nutrient recycling decreases the requirement for external feed inputs, cutting material costs and increasing overall system efficiency. However, before microalgae-assisted aquaculture can go from lab research to industrial use, issues such as cost-effective harvesting methods and safety concerns about bacterial contamination must be addressed. Once these concerns are overcome, widespread adoption of microalgae technology will significantly contribute to aquaculture's environmental and economic sustainability, as well as global sustainability in general.

12.1.2 Energy and Cost Consideration

Economic restrictions on employing microalgae for wastewater treatment have been highlighted by different researchers, particularly in terms of energy and cost. One of the key issues addressed is the high cost of carbon dioxide (CO_2) extraction and transportation, which is a critical stage in microalgae cultivation. The CO_2 compression process needs a significant amount of energy, which raises transportation costs and contributes to high manufacturing costs. Mixing CO_2 with sodium bicarbonate ($NaHCO_3$) can lower expenses, although the procedure is still energy-intensive. Furthermore, the cost of collecting microalgae biomass, which can account for up to 30% of overall production expenses, is an economic constraint. Because of their small size and density, microalgae require a large amount of energy to remove from wastewater. Furthermore, issues such as pollution in open systems and the requirement for precise nutrient balances in diverse wastewater types complicate the process and increase its overall cost. The lack of detailed economic evaluations and limited research on the financial feasibility of large-scale microalgae wastewater treatment emphasizes these economic problems [126, 127].

12.2 Environmental and Ecological Consideration

12.2.1 Potential Ecological Impacts of Large-Scale Microalgae Cultivation

The large-scale production of microalgae for wastewater treatment has both ecological advantages and disadvantages. One of the most pressing ecological concerns is the water quality and consumption necessary for such systems, which differ depending on the microalgae strain. Although low-grade wastewater, such as industrial and home effluents, might help relieve pressure on freshwater resources, it can also offer dangers if not adequately managed. Approximately 75% of worldwide wastewater is released untreated, causing severe damage to aquatic ecosystems and human health [128]. Microalgae, recognized for their adaptability and nutrient uptake efficiency, can absorb large amounts of nutrients such as carbon, nitrogen, and phosphorus from wastewater, promoting their growth. However, excessive nutrient release, particularly nitrogen and phosphorus, can cause eutrophication, which depletes oxygen in aquatic ecosystems and threatens biodiversity by creating hypoxic conditions [129]. While microalgae can help with phosphorus recovery, a vital resource controlled by a few countries, excess phosphorus in freshwater ecosystems can cause hazardous algal blooms and disturb aquatic food chains [130]. Furthermore, while nitrogen recovery by microalgae may mitigate the detrimental effects of high nitrogen levels in wastewater, poor management can result in toxic algal blooms that release ammonia and limit oxygen levels, hurting aquatic species such as fish [111]. Furthermore, the introduction of microalgae production near vulnerable ecosystems such as coastal areas or lakes may unintentionally trigger large-scale eutrophication events, affecting local biodiversity and ecosystem stability. This is especially true if non-native or genetically engineered algae get into natural water bodies, possibly outcompeting local species and disrupting food webs. Furthermore, microalgae can absorb heavy metals, herbicides, and other industrial contaminants in wastewater streams, creating bioaccumulation hazards in the food chain or during biofuel production [131]. Furthermore, viruses, pathogens, and parasites in wastewater can alter microalgae populations, causing crop losses or demanding additional treatment processes to assure the algae are safe for future use [132].

Another issue with large-scale microalgae production is its water footprint. Open pond systems, which are commonly used for microalgae farming, can result in significant water evaporation rates, especially in arid climates. Although closed photobioreactors use less water than terrestrial biofuels, they require more resources to build and operate [133]. In some situations, adding water into desert places for microalgae culture might unintentionally change local climates by boosting humidity and warmth, attracting new fauna, and disrupting the ecological balance [134]. Furthermore, while microalgae can grow on marginal soil, large-scale operations require significant amounts of land, possibly displacing agricultural or natural ecosystems, resulting in land use changes and related greenhouse gas emissions [135]. As a result, despite the environmental benefits of using microalgae for wastewater treatment, such as nutrient recovery and biofuel production, the potential ecological consequences—such as eutrophication, biodiversity loss, and water footprint, highlight the importance of careful management and system optimization.

12.2.2 Risk of Contamination and Invasive Species

Large-scale microalgae farming, especially in open pond systems, poses substantial ecological dangers, particularly in terms of contamination and the potential introduction of invasive species. Open ponds, while inexpensive, are particularly prone to pollution from outside sources. Native species can infiltrate the culture, lowering production and perhaps introducing unwelcome biological agents, such as bacteria, viruses, and other algae, which could disturb the cultivation process. This problem can be minimized by changing pond conditions to limit the growth of native species, but such measures are not infallible. Furthermore, improperly built ponds may cause groundwater contamination by leaching, especially if marine algae are grown on land, resulting in salinization or the spread of contaminants. For example, when microalgae are utilized in wastewater treatment, the potential of toxicants seeping into groundwater increases dramatically, endangering nearby ecosystems. Photobioreactors (PBRs) provide a closed environment, lowering pollution hazards; nonetheless, any potential leakage from these systems can still have a significant influence on neighboring natural water sources. Construction of large-scale ponds or reactors may potentially disturb local ecosystems, displacing native species and altering natural environments. Large-scale infrastructure projects, such as reservoirs, have demonstrated that building can result in habitat degradation, species loss, and ecosystem fragmentation, all of which reduce biodiversity. Furthermore, human activities related to pond maintenance, such as mechanical operation and increased traffic, may upset local animals, fragmenting and deteriorating natural habitats. Invasive species or hazardous illnesses may also spread from the cultivation site into nearby habitats, particularly if accidentally released, thereby causing larger ecological consequences. While environmental impact studies can help predict and avoid some of these dangers, large-scale microalgae farming still poses complicated challenges to maintaining ecological balance and biodiversity.

13 Conclusions

Developing nations are confronted with insecurity, gender inequality, unstable economies, food scarcity, and unemployment. These issues can be addressed with adequate provision of clean water and sanitation. Providing clean water and sanitation improves public health and mitigate diseases. Improved health leads to a stronger, and more stable society, reducing the potential for conflicts over resources. These improve the spate of insecurity, unstable economies, food scarcity, and unemployment. Similarly, improved health reduces the incidence of maternal mortality, with opportunities for gender equality. Clean water and sanitation can be achieved using microalgae. Microalgae contributes sustainably to clean water and sanitation. Clean water enhances quality of living sustainably. Sustaining clean water and sanitation is crucial to meeting the SDG 2030 timeline. Thus, SDG 6 relates to developmental impact in low- and middle-income countries.

References

1. Sharma R, Katoch OR (2022) Analysis of the targets and progress toward meeting the 2030 sustainable development goal-6 on clean water and sanitation: evidence from India. South Asian J Soc Stud Econ 15(3):16–26
2. Mio C, Panfilo S, Blundo B (2020) Sustainable development goals and the strategic role of business: a systematic literature review. Bus Strateg Environ 29(8):3220–3245

3. McMichael C (2019) Water, sanitation and hygiene (WASH) in schools in low-income countries: a review of evidence of impact. Int J Environ Res Public Health 16(3):359
4. Amenorfenyo DK, Huang X, Zhang Y, Zeng Q, Zhang N, Ren J, Huang Q (2019) Microalgae brewery wastewater treatment: potentials, benefits and the challenges. Int J Environ Res Public Health 16(11):1910
5. Roy A, Pramanick K (2019) Analysing progress of sustainable development goal 6 in India: past, present, and future. J Environ Manage 232:1049–1065
6. Delrue F, Álvarez-Díaz PD, Fon-Sing S, Fleury G, Sassi J-F (2016) The environmental biorefinery: using microalgae to remediate wastewater, a win-win paradigm. Energies 9(3):132
7. Scheyvens R, Banks G, Hughes E (2016) The private sector and the SDGs: the need to move beyond 'business as usual.' Sustain Dev 24(6):371–382
8. Nhemachena C, Nhamo L, Matchaya G, Nhemachena CR, Muchara B, Karuaihe ST, Mpandeli S (2020) Climate change impacts on water and agriculture sectors in Southern Africa: threats and opportunities for sustainable development. Water 12(10):2673
9. Becher O, Pant R, Verschuur J, Mandal A, Paltan H, Lawless M, Raven E, Hall J (2023) A multi-hazard risk framework to stress-test water supply systems to climate-related disruptions. Earth's Futur 11(1):e2022EF002946
10. Liang X, Zhang R, Liu C, Liu H (2018) Quantitative measurement of the sustainable water resource development system in China inspired by dissipative structure theory. Sustainability 10(11):3996
11. Yehia HMAS, Said SM (2023) Improving the quality of drinking water by raising the pH levels using a natural Na_2 SiO_3 physical field
12. Nandi A, Megiddo I, Ashok A, Verma A, Laxminarayan R (2017) Reduced burden of childhood diarrheal diseases through increased access to water and sanitation in India: a modeling analysis. Soc Sci Med 180:181–192
13. Wada Y, Bierkens MF (2014) Sustainability of global water use: past reconstruction and future projections. Environ Res Lett 9(10):104003
14. Sahoo S, Goswami S (2024) Theoretical framework for assessing the economic and environmental impact of water pollution: a detailed study on sustainable development of India. J Futur Sustain 4(1):23–34
15. D'Agostino M, Hein C (2024) Design-based solutions for water challenges: the value case approach. Blue Papers 3(1):80–89
16. Iortyom ET, Kargbo P (2023) Food and water security in the context of sustainable development. Eur J Dev Stud 3(2):10–16
17. Glavan M, Cvejić R, Tratnik M, Pintar M (2013) Geospatial analysis of water resources for sustainable agricultural water use in Slovenia. In: Current perspectives in contaminant hydrology and water resources sustainability; InTech: Rijeka, Croatia, pp 199–219
18. Yaakob MA, Mohamed RMSR, Al-Gheethi A, Aswathnarayana Gokare R, Ambati RR (2021) Influence of nitrogen and phosphorus on microalgal growth, biomass, lipid, and fatty acid production: an overview. Cells 10(2):393
19. Vadiveloo A, Foster L, Kwambai C, Bahri PA, Moheimani NR (2021) Microalgae cultivation for the treatment of anaerobically digested municipal centrate (ADMC) and anaerobically digested abattoir effluent (ADAE). Sci Total Environ 775:145853
20. Çakirsoy I, Miyamoto T, Ohtake N (2022) Physiology of microalgae and their application to sustainable agriculture: a mini-review. Front Plant Sci 13:1005991
21. Lin H-Y, Lin H-J (2018) Polyamines in microalgae: something borrowed, something new. Mar Drugs 17(1):1
22. Naselli-Flores L, Padisák J (2023) Ecosystem services provided by marine and freshwater phytoplankton. Hydrobiologia 850(12):2691–2706
23. Gan T, Zhao N, Yin G, Tan X, Wang Y (2023) Construction of a new response index for sensitive detection of the toxicity of photosynthetic inhibitory herbicides to photosynthesis of chlorella pyrenoidosa based on change characteristics of chlorophyll fluorescence rise kinetics curve. Pol J Environ Stud 32(3)

24. Mujtaba G, Lee K (2016) Advanced treatment of wastewater using symbiotic co-culture of microalgae and bacteria. Applied. Chem Eng 27(1):1–9
25. Canino A, Lemonnier C, Alric B, Bouchez A, Domaizon I, Laplace-Treyture C, Rimet F (2023) Which barcode to decipher freshwater microalgal assemblages? Tests on mock communities. Int J Limnol 59:8
26. Natrah FM, Bossier P, Sorgeloos P, Yusoff FM, Defoirdt T (2014) Significance of microalgal–bacterial interactions for aquaculture. Rev Aquac 6(1):48–61
27. Purwati N, Jayanti ET, Jannah M (2022) Community structure of freshwater microalgae in Jangkok River, Lombok Indonesia
28. Ravina M, Galletta S, Dagbetin A, Kamaleldin OAH, Mngombe M, Mnyenyembe L, Shanko A, Zanetti M (2021) Urban wastewater treatment in African countries: evidence from the hydroaid initiative. Sustainability 13(22):12828
29. Bashir I, Lone FA, Bhat RA, Mir SA, Dar ZA, Dar SA (2020) Concerns and threats of contamination on aquatic ecosystems. Bioremediation Biotechnol: Sustain Approaches Pollut Degrad 1–26
30. Bhatt P, Bhandari G, Bhatt K, Simsek H (2022) Microalgae-based removal of pollutants from wastewaters: occurrence, toxicity and circular economy. Chemosphere 306:135576
31. Ahmed SF, Mofijur M, Parisa TA, Islam N, Kusumo F, Inayat A, Badruddin IA, Khan TY, Ong HC (2022) Progress and challenges of contaminate removal from wastewater using microalgae biomass. Chemosphere 286:131656
32. Wang Z, Lee Y-Y, Scherr D, Senger RS, Li Y, He Z (2020) Mitigating nutrient accumulation with microalgal growth towards enhanced nutrient removal and biomass production in an osmotic photobioreactor. Water Res 182:116038
33. Plöhn M, Spain O, Sirin S, Silva M, Escudero-Oñate C, Ferrando-Climent L, Allahverdiyeva Y, Funk C (2021) Wastewater treatment by microalgae. Physiol Plant 173(2):568–578
34. Kumari R, Bhatnagar S, Kalra C (2022) Nitrogen assimilation in plants. In: Advances in plant nitrogen metabolism. CRC Press, pp 38–54
35. Chowdury KH, Nahar N, Deb UK (2020) The growth factors involved in microalgae cultivation for biofuel production: a review. Comput Water Energy Environ Eng 9(4):185–215
36. Abdelfattah A, Ali SS, Ramadan H, El-Aswar EI, Eltawab R, Ho S-H, Elsamahy T, Li S, El-Sheekh MM, Schagerl M (2023) Microalgae-based wastewater treatment: mechanisms, challenges, recent advances, and future prospects. Environ Sci Ecotechnology 13:100205
37. Danouche M, El Ghachtouli N, El Arroussi H (2021) Phycoremediation mechanisms of heavy metals using living green microalgae: physicochemical and molecular approaches for enhancing selectivity and removal capacity. Heliyon, 7(7)
38. Leong YK, Chang J-S (2020) Bioremediation of heavy metals using microalgae: recent advances and mechanisms. Biores Technol 303:122886
39. Nguyen HT, Yoon Y, Ngo HH, Jang A (2021) The application of microalgae in removing organic micropollutants in wastewater. Crit Rev Environ Sci Technol 51(12):1187–1220
40. Molinuevo-Salces B, Riaño B, Hernández D, Cruz García-González M (2019) Microalgae and wastewater treatment: advantages and disadvantages. In: Microalgae biotechnology for development of biofuel and wastewater treatment, pp 505–533
41. Udaiyappan AFM, Hasan HA, Takriff MS, Abdullah SRS (2017) A review of the potentials, challenges and current status of microalgae biomass applications in industrial wastewater treatment. J Water Process Eng 20:8–21
42. Thanigaivel S, Priya A, Gnanasekaran L, Hoang TK, Rajendran S, Soto-Moscoso M (2022) Sustainable applicability and environmental impact of wastewater treatment by emerging nanobiotechnological approach: future strategy for efficient removal of contaminants and water purification. Sustain Energy Technol Assess 53:102484
43. Alkhadra MA, Su X, Suss ME, Tian H, Guyes EN, Shocron AN, Conforti KM, De Souza JP, Kim N, Tedesco M (2022) Electrochemical methods for water purification, ion separations, and energy conversion. Chem Rev 122(16):13547–13635
44. Avila R, Justo Á, Carrero E, Crivillés E, Vicent T, Blánquez P (2022) Water resource recovery coupling microalgae wastewater treatment and sludge co-digestion for bio-wastes valorisation at industrial pilot-scale. Biores Technol 343:126080

45. Hussain F, Shah SZ, Ahmad H, Abubshait SA, Abubshait HA, Laref A, Manikandan A, Kusuma HS, Iqbal M (2021) Microalgae an ecofriendly and sustainable wastewater treatment option: biomass application in biofuel and bio-fertilizer production. A review. Renew Sustain Energy Rev 137:110603
46. Reisoglu Ş, Aydin S (2023) Microalgae as a promising candidate for fighting climate change and biodiversity loss
47. Ali S, Peter AP, Chew KW, Munawaroh HSH, Show PL (2021) Resource recovery from industrial effluents through the cultivation of microalgae: a review. Biores Technol 337:125461
48. Sauvé S, Lamontagne S, Dupras J, Stahel W (2021) Circular economy of water: tackling quantity, quality and footprint of water. Environ Dev 39:100651
49. Castellet-Viciano L, Hernández-Chover V, Hernández-Sancho F (2022) The benefits of circular economy strategies in urban water facilities. Sci Total Environ 844:157172
50. Giakoumis T, Vaghela C, Voulvoulis N (2020). The role of water reuse in the circular economy. In: Advances in chemical pollution, environmental management and protection, vol 5. Elsevier, pp 227–252
51. Hu J, Nagarajan D, Zhang Q, Chang J-S, Lee D-J (2018) Heterotrophic cultivation of microalgae for pigment production: a review. Biotechnol Adv 36(1):54–67
52. Mohan SV, Nikhil GN, Chiranjeevi P, Reddy CN, Rohit MV, Kumar AN, Sarkar O (2016) Waste biorefinery models towards sustainable circular bioeconomy: critical review and future perspectives. Biores Technol 215:2–12
53. Shahzad MW, Burhan M, Ybyraiymkul D, Ng KC (2019) Desalination processes' efficiency and future roadmap. Entropy 21(1):84
54. Taheri R, Razmjou A, Szekely G, Hou J, Ghezelbash GR (2016) Biodesalination—on harnessing the potential of nature's desalination processes. Bioinspir Biomim 11(4):041001
55. Kumar Patel A, Tseng Y-S, Rani Singhania R, Chen C-W, Chang J-S, Di Dong C (2021) Novel application of microalgae platform for biodesalination process: a review. Bioresour Technol 337:125343. https://doi.org/10.1016/j.biortech.2021.125343
56. Azmi AS, Sani FS, Ali F, Mel M (2018) Interactive effect of temperature, phand light intensity on biodesalination of seawater by synechococcus sp. PCC 7002 and on the cyanobacteria growth. J Adv Res Fluid Mech Therm Sci 52(1):85–93
57. Figler A, B-Béres V, Dobronoki D, Márton K, Nagy SA, Bácsi I (2019) Salt tolerance and desalination abilities of nine common green microalgae isolates. Water 11(12):2527
58. Patel AK, Joun J, Sim SJ (2020) A sustainable mixotrophic microalgae cultivation from dairy wastes for carbon credit, bioremediation and lucrative biofuels. Biores Technol 313:123681
59. Dahlin LR, Gerritsen AT, Henard CA, Van Wychen S, Linger JG, Kunde Y, Guarnieri MT (2019) Development of a high-productivity, halophilic, thermotolerant microalga Picochlorum renovo. Commun Biol 2(1):388
60. Pal P, Khairnar K, Paunikar WN (2014) Causes and remedies for filamentous foaming in activated sludge treatment plant. Global Nest J 16(4):762–772
61. Bedia J, Belver C (2022) 11—Carbon materials from sewage sludge. In: Shah MP, Rodriguez-Couto S, Shah N, Banerjee R (eds) Development in waste water treatment research and processes. Elsevier, pp 175–196. https://doi.org/10.1016/B978-0-323-85584-6.00012-1
62. Yahampath Arachchige Don CDY, Babel S (2020) Comparing the performance of microbial fuel cell with mechanical aeration and photosynthetic aeration in the cathode chamber. Int J Hydrog Energy. https://doi.org/10.1016/j.ijhydene.2020.09.068
63. Abudaqqa WS, Madhuranthakam CMR, Chaalal O (2024) Algae-based membrane bioreactors: a mini review on their progress and processes for wastewater treatment. Journal of Water Process Engineering 59:104937
64. Onyeaka H, Miri T, Obileke K, Hart A, Anumudu C, Al-Sharify ZT (2021) Minimizing carbon footprint via microalgae as a biological capture. Carbon Capture Sci Technol 1:100007. https://doi.org/10.1016/j.ccst.2021.100007
65. Al-Asheh S, Bagheri M, Aidan A (2021) Membrane bioreactor for wastewater treatment: a review. Case Stud Chem Environ Eng 4:100109. https://doi.org/10.1016/j.cscee.2021.100109

66. Sun F, Wang X, Li X (2013) An innovative membrane bioreactor (MBR) system for simultaneous nitrogen and phosphorus removal. Process Biochem 48:1749–1756. https://doi.org/10.1016/j.procbio.2013.08.009
67. Abdel-Raouf N, Al-Homaidan AA, Ibraheem IBM (2012) Microalgae and waste water treatment. Saudi J Biol Sci 19(3):257–275. https://doi.org/10.1016/j.sjbs.2012.04.005
68. Mofijur M, Hasan MM, Sultana S, Kabir Z, Djavanroodi F, Ahmed SF, Khan TY (2023) Advancements in algal membrane bioreactors: Overcoming obstacles and harnessing potential for eliminating hazardous pollutants from wastewater. Chemosphere 336:139291
69. Sun L, Tian Y, Zhang J, Cui H, Zuo W, Li J (2018) A novel symbiotic system combining algae and sludge membrane bioreactor technology for wastewater treatment and membrane fouling mitigation: Performance and mechanism. Chem Eng J 344:246–253
70. Venault A, Ballad MRB, Huang Y-T, Liu Y-H, Kao C-H, Chang Y (2016) Antifouling PVDF membrane prepared by VIPS for microalgae harvesting. Chem Eng Sci 142:97–111
71. Jiang L, Liu Y, Guo F, Zhou Z, Jiang J, You Z, Wu Z (2021) Evaluation of nutrient removal performance and resource recovery potential of anaerobic/anoxic/aerobic membrane bioreactor with limited aeration. Biores Technol 340:125728. https://doi.org/10.1016/j.biortech.2021.125728
72. Mortezaeikia V, Tavakoli O, Yegani R, Faramarzi MA (2014) Carbon dioxide biofixation using microalgae in hollow fiber membrane photo-bioreactor (HFMPB). Retrieved from https://www.sid.ir/FileServer/SE/219e201408773.pdf
73. Ewusi-Mensah D, Huang J, Chaparro LK, Rodenas P, Ramírez-Moreno M, Ortiz JM, Esteve-Núñez A (2021) Algae-assisted microbial desalination cell: analysis of cathode performance and desalination efficiency assessment. Processes 9(11):2011
74. Jingyu H, Ewusi-Mensah D, Norgbey E (2017) Microbial desalination cells technology: a review of the factors affecting the process, performance and efficiency. Desalin Water Treat 87:140–159
75. Ucar D, Zhang Y, Angelidaki I (2017) An overview of electron acceptors in microbial fuel cells. Front Microbiol 8:643
76. Ebrahimi A, Najafpour GD, Kebria DY (2018) Performance of microbial desalination cell for salt removal and energy generation using different catholyte solutions. Desalination 432:1–9
77. Arana TJ, Gude VG (2018) A microbial desalination process with microalgae biocathode using sodium bicarbonate as an inorganic carbon source. Int Biodeterior Biodegradation 130:91–97
78. Olguín EJ, Sánchez-Galván G (2012) Heavy metal removal in phytofiltration and phycoremediation: the need to differentiate between bioadsorption and bioaccumulation. New Biotechnol 30(1):3–8
79. Pathak VV, Kothari R, Chopra AK, Singh DP (2015) Experimental and kinetic studies for phycoremediation and dye removal by Chlorella pyrenoidosa from textile wastewater. J Environ Manage 163:270–277
80. Hernández D, Solana M, Riaño B, García-González MC, Bertucco A (2014) Biofuels from microalgae: lipid extraction and methane production from the residual biomass in a biorefinery approach. Biores Technol 170:370–378
81. Bhatt NC, Panwar A, Bisht TS, Tamta S (2014) Coupling of algal biofuel production with wastewater. Sci World J 2014:1–10. https://doi.org/10.1155/2014/210504
82. Pal M, Yesankar PJ, Dwivedi A, Qureshi A (2020) Biotic control of harmful algal blooms (HABs): a brief review. J Environ Manage 268:110687
83. Li S, Tao Y, Zhan X-M, Dao G-H, Hu H-Y (2020) UV-C irradiation for harmful algal blooms control: a literature review on effectiveness, mechanisms, influencing factors and facilities. Sci Total Environ 723:137986
84. Oladoja NA, Ali J, Lei W, Yudong N, Pan G (2020) Tapping into the ballast potential of sparingly soluble salts for enhanced floc physiognomies in algae biomass harvesting. Sep Purif Technol 248:117150
85. Teixeira MR, Sousa V, Rosa MJ (2010) Investigating dissolved air flotation performance with cyanobacterial cells and filaments. Water Res 44(11):3337–3344

86. Jiang S, Xiao S, Chu H, Zhao F, Yu Z, Zhou X, Zhang Y (2020) Intelligent mitigation of fouling by means of membrane vibration for algae separation: dynamics model, comprehensive evaluation, and critical vibration frequency. Water Res 182:115972. https://doi.org/10.1016/j.watres.2020.115972
87. Park J, Church J, Son Y, Kim K-T, Lee WH (2017) Recent advances in ultrasonic treatment: challenges and field applications for controlling harmful algal blooms (HABs). Ultrason Sonochem 38:326–334
88. Kinley CM, Iwinski KJ, Hendrikse M, Geer TD, Rodgers JH Jr (2017) Cell density dependence of microcystis aeruginosa responses to copper algaecide concentrations: implications for microcystin-LR release. Ecotoxicol Environ Saf 145:591–596
89. Zhou S, Shao Y, Gao N, Zhu S, Li L, Deng J, Zhu M (2014) Removal of Microcystis aeruginosa by potassium ferrate (VI): impacts on cells integrity, intracellular organic matter release and disinfection by-products formation. Chem Eng J 251:304–309
90. Tsai K-P (2016) Management of target algae by using copper-based algaecides: effects of algal cell density and sensitivity to copper. Water Air Soil Pollut 227(7):238. https://doi.org/10.1007/s11270-016-2926-8
91. Hanotu J, Bandulasena HCH, Zimmerman WB (2012) Microflotation performance for algal separation. Biotechnol Bioeng 109(7):1663–1673. https://doi.org/10.1002/bit.24449
92. Christenson L, Sims R (2011) Production and harvesting of microalgae for wastewater treatment, biofuels, and bioproducts. Biotechnol Adv 29(6):686–702
93. Granados MR, Acién FG, Gómez C, Fernández-Sevilla JM, Grima EM (2012) Evaluation of flocculants for the recovery of freshwater microalgae. Biores Technol 118:102–110
94. Li Y, Lei X, Zhu H, Zhang H, Guan C, Chen Z, Zheng T (2016) Chitinase producing bacteria with direct algicidal activity on marine diatoms. Sci Rep 6(1):21984
95. Lu K, Jin C, Dong S, Gu B, Bowen SH (2006) Feeding and control of blue-green algal blooms by tilapia (Oreochromis Niloticus). Hydrobiologia 568(1):111–120. https://doi.org/10.1007/s10750-006-0023-5
96. Li B, Yin Y, Kang L, Feng L, Liu Y, Du Z, Zhang L (2021) A review: Application of allelochemicals in water ecological restoration——algal inhibition. Chemosphere 267:128869. https://doi.org/10.1016/j.chemosphere.2020.128869
97. Hong Y, Hu H-Y, Xie X, Sakoda A, Sagehashi M, Li F-M (2009) Gramine-induced growth inhibition, oxidative damage and antioxidant responses in freshwater cyanobacterium microcystis aeruginosa. Aquat Toxicol 91(3):262–269
98. Hu Z, Li D, Guan D (2020) Water quality retrieval and algae inhibition from eutrophic freshwaters with iron-rich substrate based ecological floating beds treatment. Sci Total Environ 712:135584
99. Song J, Wang X, Ma J, Wang X, Wang J, Xia S, Zhao J (2018) Removal of microcystis aeruginosa and microcystin-LR using a graphitic-C3N4/TiO2 floating photocatalyst under visible light irradiation. Chem Eng J 348:380–388
100. Bockstaller C, Guichard L, Keichinger O, Girardin P, Galan M-B, Gaillard G (2009) Comparison of methods to assess the sustainability of agricultural systems. A review. Agron Sustain Dev 29(1):223–235. https://doi.org/10.1051/agro:2008058
101. Schwarzenbach RP, Egli T, Hofstetter TB, Von Gunten U, Wehrli B (2010) Global water pollution and human health. Annu Rev Environ Resour 35(1):109–136. https://doi.org/10.1146/annurev-environ-100809-125342
102. Rout PR, Zhang TC, Bhunia P, Surampalli RY (2021) Treatment technologies for emerging contaminants in wastewater treatment plants: a review. Sci Total Environ 753:141990
103. Crini G, Lichtfouse E (2019) Advantages and disadvantages of techniques used for wastewater treatment. Environ Chem Lett 17(1):145–155. https://doi.org/10.1007/s10311-018-0785-9
104. Acién FG, Gómez-Serrano C, Morales-Amaral MM, Fernández-Sevilla JM, Molina-Grima E (2016) Wastewater treatment using microalgae: how realistic a contribution might it be to significant urban wastewater treatment? Appl Microbiol Biotechnol 100(21):9013–9022. https://doi.org/10.1007/s00253-016-7835-7

105. Michalak I, Chojnacka K, Witek-Krowiak A (2013) State of the art for the biosorption process—a review. Appl Biochem Biotechnol 170(6):1389–1416. https://doi.org/10.1007/s12010-013-0269-0
106. Ubando AT, Africa ADM, Maniquiz-Redillas MC, Culaba AB, Chen W-H, Chang J-S (2021) Microalgal biosorption of heavy metals: a comprehensive bibliometric review. J Hazard Mater 402:123431
107. Wang B, Liang W, Guo Z, Liu W (2015) Biomimetic super-lyophobic and super-lyophilic materials applied for oil/water separation: A new strategy beyond nature. Chem Soc Rev 44(1):336–361
108. Ferro L, Gentili FG, Funk C (2018) Isolation and characterization of microalgal strains for biomass production and wastewater reclamation in Northern Sweden. Algal Res 32:44–53
109. Wang L, Min M, Li Y, Chen P, Chen Y, Liu Y, Ruan R (2010) Cultivation of green algae chlorella sp. in different wastewaters from municipal wastewater treatment plant. Appl Biochem Biotechnol 162(4):1174–1186. https://doi.org/10.1007/s12010-009-8866-7
110. Masojídek J, Torzillo G (2014) Mass cultivation of freshwater microalgae
111. Wu LF, Chen PC, Huang AP, Lee CM (2012) The feasibility of biodiesel production by microalgae using industrial wastewater. Biores Technol 113:14–18
112. Prandini JM, Da Silva MLB, Mezzari MP, Pirolli M, Michelon W, Soares HM (2016) Enhancement of nutrient removal from swine wastewater digestate coupled to biogas purification by microalgae Scenedesmus spp. Biores Technol 202:67–75
113. Borowitzka MA (2018) Microalgae in medicine and human health: a historical perspective. In: Microalgae in health and disease prevention. Elsevier, pp 195–210. Retrieved from https://www.sciencedirect.com/science/article/pii/B9780128114056000098
114. Novoveská L, Zapata AK, Zabolotney JB, Atwood MC, Sundstrom ER (2016) Optimizing microalgae cultivation and wastewater treatment in large-scale offshore photobioreactors. Algal Res 18:86–94
115. Wollmann F, Dietze S, Ackermann J, Bley T, Walther T, Steingroewer J, Krujatz F (2019) Microalgae wastewater treatment: Biological and technological approaches. Eng Life Sci 19(12):860–871. https://doi.org/10.1002/elsc.201900071
116. Le Truong G, Tran D-T, Do V, Nguyen V (2019). Design considerations of microalgal culture ponds and photobioreactors for wastewater treatment and biomass cogeneration. https://doi.org/10.1007/978-981-13-2264-8_21
117. Fortier M-OP, Roberts GW, Stagg-Williams SM, Sturm BS (2017) Determination of the life cycle climate change impacts of land use and albedo change in algal biofuel production. Algal Res 28:270–281
118. Handler RM, Shi R, Shonnard DR (2017) Land use change implications for large-scale cultivation of algae feedstocks in the United States Gulf Coast. J Clean Prod 153:15–25
119. Searchinger T, Heimlich R, Houghton RA, Dong F, Elobeid A, Fabiosa J, Yu T-H (2008) Use of U.S. croplands for biofuels increases greenhouse gases through emissions from land-use change. Science 319(5867):1238–1240. https://doi.org/10.1126/science.1151861
120. Posadas E, Alcántara C, García-Encina PA, Gouveia L, Guieysse B, Norvill Z, Koreiviene J (2017) Microalgae cultivation in wastewater. In: Microalgae-based biofuels and bioproducts. https://www.sciencedirect.com/science/article/pii/B9780081010235000030
121. Park JBK, Craggs RJ, Shilton AN (2011) Wastewater treatment high rate algal ponds for biofuel production. Biores Technol 102(1):35–42
122. De Godos I, Arbib Z, Lara E, Cano R, Muñoz R, Rogalla F (2017) Wastewater treatment in algal systems. Innov Wastewater Treat Resour Recover Technol: Impacts Energy Econ Environ 76
123. Sutherland DL, Howard-Williams C, Turnbull MH, Broady PA, Craggs RJ (2015) Enhancing microalgal photosynthesis and productivity in wastewater treatment high rate algal ponds for biofuel production. Biores Technol 184:222–229
124. Szulc J, Otlewska A, Okrasa M, Majchrzycka K, Sulyok M, Gutarowska B (2017) Microbiological contamination at workplaces in a combined heat and power (CHP) station processing plant biomass. Int J Environ Res Public Health 14(1):99. https://doi.org/10.3390/ijerph14010099

125. Luo S, Wu X, Jiang H, Yu M, Liu Y, Min A, Ruan R (2019) Edible fungi-assisted harvesting system for efficient microalgae bio-flocculation. Biores Technol 282:325–330. https://doi.org/10.1016/j.biortech.2019.03.033
126. Molazadeh M, Ahmadzadeh H, Pourianfar HR, Lyon S, Rampelotto PH (2019) The use of microalgae for coupling wastewater treatment with CO2 biofixation. Front Bioeng Biotechnol 7:42
127. Wang S, Mukhambet Y, Esakkimuthu S (2022) Integrated microalgal biorefinery–routes, energy, economic and environmental perspectives. J Clean Prod 348:131245
128. Swyngedouw E (2013) UN Water Report 2012: depoliticizing water. Dev Chang 44(3)
129. Sheehan J, Dunahay T, Benemann J, Roessler P (1998) A look back at the US department of energy's aquatic species program: biodiesel from algae. Natl Renew Energy Lab 328:1–294
130. Linderholm K, Tillman A-M, Mattsson JE (2012) Life cycle assessment of phosphorus alternatives for Swedish agriculture. Resour Conserv Recycl 66:2739
131. Subashchandrabose SR, Ramakrishnan B, Megharaj M, Venkateswarlu K, Naidu R (2013) Mixotrophic cyanobacteria and microalgae as distinctive biological agents for organic pollutant degradation. Environ Int 51:59–72
132. Letcher PM, Lopez S, Schmieder R, Lee PA, Behnke C, Powell MJ, McBride RC (2013) Characterization of Amoeboaphelidium protococcarum, an algal parasite new to the cryptomycota isolated from an outdoor algal pond used for the production of biofuel. PLoS ONE 8(2):e56232
133. Yang J, Xu M, Zhang X, Hu Q, Sommerfeld M, Chen Y (2011) Life-cycle analysis on biodiesel production from microalgae: water footprint and nutrients balance. Biores Technol 102(1):159–165
134. Sustainable Development of Algal Biofuels in the United States (2012) National Academies Press, Washington, D.C. https://doi.org/10.17226/13437
135. Venteris ER, Skaggs RL, Coleman AM, Wigmosta MS (2013) A GIS cost model to assess the availability of freshwater, seawater, and saline groundwater for algal biofuel production in the United States. Environ Sci Technol 47(9):4840–4849. https://doi.org/10.1021/es304135b

Net-Zero Emission and Microalgae Biotechnology Carbon Dioxide Sequestration

Veluru Sridevi, Husam Talib Hamzah, Avinash Terapalli, Jampana Vishnu Varma, Chinta Sankar Rao, and Potnuri Ramesh

Abstract In recent decades, people all around the globe have been paying attention to the dire situation caused by pollution and climate change, which are both caused by human activities. Although carbon capture and storage (CCS) technologies were originally thought to be a viable solution to this concerning situation, they are now recognized as economically inconvenient and the environmental consequences they may have in the future are uncertain. On the other hand, microalgae-based biological CO_2 capture is thought of as a captivating medium in order to repurpose the surplus carbon dioxide gas produced by fossil fuel combustion, vehicles, and natural disasters like volcanic eruptions, decay of organic matter, and wildfires. In addition, microalgae have the ability to collect carbon dioxide and convert it into biomass; it might then be used as a carbon source for making lipids in order to create bioenergy and other products with added value. These materials will hopefully one day serve as a more environmentally friendly alternative to transportation fuels made from petroleum in a way that doesn't interfere with or negatively impact the crops and food supply. This review focuses on biological carbon sequestration, influencing factors and methods to enhance microalgae's CO_2 sequestration and genetic engineering's prospective aims for improving CO_2 sequestration. We draw attention to difficulties with existing methods and looking forward with the intention of making a positive impact on environmental sustainability, achieving zero emissions, and valuing carbon dioxide for a circular carbon economy (CCE).

Keywords Microbial CO_2 sequestration · CCE · Net-zero emission · Genetic engineering

V. Sridevi (✉) · H. T. Hamzah · A. Terapalli · J. V. Varma
Department of Chemical Engineering, Andhra University, Visakhapatnam, Andhra Pradesh, India
e-mail: prof.vsridevi@andhrauniversity.edu.in

C. S. Rao · P. Ramesh
Department of Chemical Engineering, NITK, Surathkal, India

A. Khan and K. A. Alzahrani (eds.), *Microalgae Biotechnology and Sustainable Development*, Sustainable Materials and Technology,
https://doi.org/10.1007/978-981-95-2597-3_6

1 Introduction

An increasing number of plant and animal species, along with people, are facing extinction due to the effects of climate change. Carbon dioxide (CO_2), water vapor, methane (CH_4), nitrous oxide (N_2O), and fluorinated gases are among the greenhouse gases (GHGs) whose emissions have been steadily rising, worsening climate change [1]. The burning of fossil fuels is the main human-caused factor in the increase of greenhouse gas emissions [2]. The atmospheric CO_2 level on Earth has risen from around 313 parts per million in 1960 to around 411 parts per million now. An increase in ocean acidity due to elevated atmospheric CO_2 levels has far-reaching consequences for marine life [3, 4]. Therefore, right now, more than ever, we need a plan to bring the atmospheric CO_2 level down or at least keep it stable. For example, the Paris Agreement of 2015 [5], the Kyoto Protocol of 1997 [COP26], and numerous other international agreements have been signed by various nations in an effort to decrease emissions of greenhouse gases. There are two main ways to lower carbon dioxide emissions: (i) switching to renewable energy while simultaneously reducing reliance on fossil fuels; (ii) and sequestration of carbon using several biological, chemical, and physical procedures [6, 7]. There has been a lot of research into the physical means of reducing carbon emissions. However, current technologies have a number of economic and technological restrictions. That is why it is so important to both improve the current technology and come up with appropriate substitutes. When compared to physical and chemical approaches, biological CO_2 fixation appears to be an environmentally acceptable and cost-effective alternative. In the dark phase of photosynthesis, carbon dioxide is absorbed by photosynthetic organisms and contributes significantly to keeping atmospheric CO_2 levels stable. The efficiency and biomass productivity of CO_2 fixation were both higher in phytoplankton than in other photosynthetic species [5]. Half of the world's primary production, or around 50 gigatonnes of CO_2 fixed per year, is attributed to marine phytoplankton. Here, studies on microalgae's ability to sequester carbon dioxide have garnered interest from all over the world [8–13]. When it comes to carbon dioxide absorption, microalgae outperform vascular plants by a factor of 10 to 50 or supplying nourishment to living beings [14–16]. The carbon concentration mechanism (CCM) is unique to microalgae and allows them to absorb carbon dioxide. The thylakoid membranes are exposed to elevated CO_2 levels in this process by means of a specific organelle called a pyrenoid [17]. One of the most important enzymes in photosynthesis, ribulose-1,5-bisphosphate carboxylase/oxygenase (Rubisco) helps with carbon absorption or storage when there is more carbon dioxide around the thylakoid membrane. Due to its evolutionary history in humid, low-oxygen conditions, rubisco has a poor affinity for carbon dioxide,thus, the pyrenoid continually offers a setting for improved CO_2 fixation [18]. A greater tolerance for elevated CO_2 concentration was seen in the microalgae cultivated under different settings for carbon sequestration. Additionally, microalgae have shown promise in a number of studies for the sequestration of flue gas (NOx and SOx) [19]. Microalgae are able to withstand harsh environments and produce a high biomass yield. Thus, microalgae have the ability to capture carbon dioxide and produce biofuels [20, 21]. In addition to their usage in wastewater treatment and biomass production, microalgae have a wide range of potential uses in the manufacture of biogas, bioplastics, and fertilizer [22, 23]. Because of their great photosynthetic efficiency and low nutritional requirements, algae are easily cultivated for their many uses. Topics covered in this research include methods for enhancing microalgae's CO_2 sequestration capacity and biological carbon sequestration and possible genetic engineering targets to improve CO_2 sequestration.

2 Carbon Dioxide (CO_2) is a Strong Climatic and Global Warming Agent

In the early days of the Second Industrial Revolution, when mass production and electricity generation were defining features, energy production relied heavily on the combustion of fossil fuels. Over the last several decades, humans have released vast quantities of carbon dioxide into the atmosphere as a result of activities such as producing electricity, making cement, and running many homes on fossil fuels. Carbon dioxide emissions worldwide rose by more than 70% between 1970 and 2002, according to the research. Ever since then, people have been worried about the harsh effects of CO_2 on the ecosystem and its steadily rising atmospheric concentration [24]. To reduce the atmospheric concentration of CO_2, it is essential to capturing CO_2 from big point sources. According to Schubert and Jahren [25], while human activities do contribute to CO_2 emissions, natural processes (such as decomposition, ocean discharge, and respiration) account for a greater proportion of CO_2 emissions. Nevertheless, photosynthesis and rock weathering provide a close check on the natural CO_2 sources. Because of this equilibrium, CO_2 levels in the air were as low as 260–280 ppm for ten thousand years prior to the beginning of the industrial revolution [26]. Electricity consumption soared during the industrial revolution, and by the turn of the millennium, power generation had jumped by over half [27]. In 2002, the estimated worldwide CO_2 emissions from fossil fuels reached about 25,000 metric tons. The majority of the world's energy will continue to come from fossil fuels, which will see a growth of more than 90% [28]. This demonstrates that the demand for energy around the world has been rising and is projected to rise by more than 67% by the year 2030 [29]. This highlights the need for determining where CO_2 emissions are coming from and how to stop them. Table 1 lists potential sources of CO_2 emissions and ways to reduce them and shows that fossil fuel combustion engines and coal-fired power plants are the main contributors to the anthropogenic CO_2 emissions that have been releasing into the atmosphere. Point sources that remain in one place account for the vast majority of man-made emissions. Therefore, the best way to stop CO_2 emissions, according to this research, is carbon capture, storage, and use (CCSU). CCSUs are essentially just huge stationary point producers of CO_2 that

Table 1 Emissions of carbon dioxide and steps to reduce them from key sources

Sources	CO_2 emission (billion Mt)	Proposed preventive option	References
Anthropogenic source			
Combustion engines that use fossil fuels	392	CCSU	[31]
Cement manufacturing facilities	113	CCSU	[32]
Power generation (coal-fired power plants)	279	CCSU, integration to methanol plant	[31]
Transportation	191	Blending fuels with biomass	[33]
Industrial manufacturing	178	CCSU	[33]
Changes in land usage	13	–	[34]
Nonanthropogenic/natural sources			
Respiration in plants, animals, and humans	7	–	[35]
Hydrology and weather interaction	7	–	[35]
Soil respiration and decomposition	1.54	–	[36]
Volcanic eruptions	0.15	–	[36]

produce waste (e.g., power plants that burn coal or gas) and its disposal in a secure repository that prevents its reintroduction to the atmosphere [30].

3 Methods and Species of Microalgae for Carbon Dioxide Sequestration

3.1 Carbon Sequestration Microalgae

When it comes to photosynthesis and carbon sequestration, algae are among the most efficient organisms. Their size and structure allow them to be categorized as either macroalgae or microalgae. Environment and morphology are the primary criteria for microalgae classification. They might be heterotrophic or autotrophic, or perhaps both, depending on the circumstances. In order to grow, autotrophic microalgae need salts, inorganic compounds, and the right kind of light. In contrast, heterotrophic microalgae derive their energy from external sources, such as organic compounds and nutrients. According to Greenwell et al. [37], microalgae are often classified based on their cellular structure, color, and life cycle. Two classes of microalgae are prokaryotic, and nine classes are eukaryotic biofuels and other goods with additional value could be made possible by these creatures [38]. Microalgae are well-suited to utilize CO_2 from the exhaust systems of power plants due to their inherent ability to endure high quantities of CO_2 [39]. The first biological method to fix carbon for contemporary usage was photosynthesis. It is well-documented that several microalgal species utilize carbonates, including $NaHCO_3$ and Na_2CO_3, in order to promote the expansion of cells [40]. The amount of oil can range from 0 to 80%, depending on the species. So, microalgae may be used to store carbon dioxide and turn it into energy [41]. Microalgae have a diverse range of cellular lipids, which suggests they can thrive in carbon-rich conditions. Scientists studied how different species fared when exposed to elevated CO_2 levels, paying special attention to those that could maintain such levels while simultaneously generating abundant macromolecules like lipids and triglycerides [42]. *Many different species of algae have been investigated for their biosequestration potential, including Botryococcus braunii, Chlorella kessleri, Dunaliella tertiolecta, Isochrysis galbana, Cholorococcum littorale, Spirulina platensis, Nannochloropsis, Dunaliella salina, Scenedesmus obliquus, and Chlorella sp.* A lot of carbon capture systems use freshwater microalgae like *Chlorella, Scenedesmus,* and *Spirulina platensis* [43]. Microalgae have great promise as a CO_2 sequestration technology, according to Cheah et al. [44], which found that grown microalgae can fix 1.83 kg of CO_2 per kilogram. *Chlorella vulgaris* and *Anabaena* sp., for example, may fix CO_2 at rates of 6.24 g/L/d, respectively [45].

A great algal species can withstand high concentrations of CO_2, temperatures, pH levels, nutrient concentrations, and pollutants; additionally, its bio sequestration rate is particularly high. The microalgae species' CO2 tolerance is considered throughout the selection process. *Scenedesmus* sp. is able to develop in environments with CO_2 concentrations between 10 and 20% (v/v), as stated in a paper, despite the fact that a mere 2% (v/v) of CO_2 is ideal. There is a strong relationship between the microalgal species' exposure to CO2 and their biomass production [46]. The microalgae Chlorella sp. can thrive in conditions of 30 °C, pH 5–6, and 40% CO_2 (Chen et al. 2014). The growth rate of *Nannochloropsis* sp. increased from 0.33 to 0.52 per day, or 58% quicker, when cultivated with 15% (v/v) CO_2 [47]. Some species of microalgae cannot grow in environments with CO_2 concentrations higher than 5% (v/v).

4 Microalgae Carbon Sequestration Factor

4.1 CO_2 Concentration

For microalgae to grow and develop, they need carbon dioxide so they can create their carbon skeleton through photosynthesis. Transmutation from a gas to a liquid is the process that fixes carbon dioxide. The tolerance of different species to carbon dioxide concentrations varies. A study found that *Chlorella vulgaris* grew best at a CO_2 concentration of 5% (v/v) and that a concentration of 15% (v/v) inhibited growth. In most cases, when CO_2 concentrations are high, microalgae are unable to grow because the chloroplast stroma region becomes acidified, rendering inactive the Calvin-Benson cycle's essential enzymes [48, 49]. According to a comprehensive analysis of how CO_2 concentration affects different types of algae, there is no sweet spot for CO_2 levels that would maximize biomass [50]. Some of the microalgae studied, including *Nannochloropsis oculata, Scenedesmus obliquus,* and *Chlorella vulgaris,* demonstrated an impressive level of carbon dioxide tolerance. Singh and Singh [50] observed that microalgae *Chlorella* sp. T-1, *Scenedesmus* sp., and *Euglena gracilis* exhibited excellent tolerance to varying concentrations of carbon dioxide (i.e., 100%, 80%, and 45%, respectively), but that 10%, 10–20%, and 5% of carbon dioxide produced the highest biomass productivity.

4.2 pH

The effectiveness of carbon sequestration, photosynthetic activity, and nutrient uptake are all significantly impacted by the pH of the cultivation medium. With the exception of certain *cyanobacteria,* most microalgal cultures thrive in pH 6.5–8.5. On the other hand, high-pH environments are ideal for CO_2 absorption because they promote the development of algae that can withstand elevated concentrations of the gas via enriching the culture media with more free CO_2 [51]. A new research found that pH has no substantial influence on carbon absorption at pH 10.6–7.0 and carbon concentrations <9.52 mmol/L, pH significantly affects carbon absorption at concentrations greater than the same [52].

4.3 Temperature

To a photosynthetic organism, temperature is a key component in growth and development. Temperatures between 0 and 30 °C are ideal for microalgae growth, while they can be observed to grow in a wide range (5 to 40 °C) [53]. Different species have different ideal temperatures,for example, *Chlorella vulgaris* prefers 30 °C and *Nannochloropsis oculata* 20 °C [54]. Algae are unable to sequester carbon when temperatures are low because the enzymes ribulose-1,5-bisphosphate (Rubisco) and carboxylase are less active. Both photosynthesis and photorespiration rely on the enzyme. Microalgae also experience changes in metabolic activity and photorespiration when exposed to extremely high temperatures. Above 40 °C, microalgae typically experience temperature stress, which renders their oxygen-evolving ability inactive and causes charge separation of PSII. A change in the 33-kDa protein's structure renders the oxygen evolution activity inactive [55]. When it comes to dissolving

CO_2 in a liquid, temperature is inversely related,as the temperature raises, the CO_2 solubility in the media for growing algae drops.

4.4 Irradiance

Light is essential for photoautotrophic and mixotrophic microalgae to bioconvert CO_2. The photosynthetic reaction consists of two steps: the light-dependent step and the light-independent step. A process that is dependent on light converts molecules of photons into the biological substances ATP and NADPH. Its role in the Calvin-Benson cycle involves fixing or sequestering carbon when Rubisco is present [56]. Carbon sequestration (Calvin-Benson cycle) in microalgae requires optimal irradiation [57].

5 Genetic Engineering

At the molecular level, genetic engineering modifies the target genome by modifying gene expression and transcription through the use of recombinant DNA. Three possible routes exist in theory: (1) in order to establish the required phenotype, introduce heterologous genes; (2) gene disruption, gene deletion, or RNA interference can be used to halt gene expression or reduce the degree of expression of already-existing genes; (3) in order to boost and induce expression using promoters that are not native. The method is difficult, but it's worth it to enhance microalgal biomass, promote photosynthesis, and increase CO_2 absorption rates [58]. Overexpression of a nuclear-encoded gene in *Nannochloropsis oceanica* increases biomass production. There was a 32% increase in the mutant growth rate and a 46% increase in biomass accumulation at CO_2 concentrations in the air [59]. As the global population continues to soar, one important way that genetic engineering and biotechnological technologies can help mitigate climate change is by reducing the amount of greenhouse gas emissions through the use of GMO crops for biofuels and sustainable crop production [60].

6 Possible Areas to Improve CO_2 Sequestration Using Genetic Engineering

To enhance microalgae CO_2 fixation, three primary target sites should be investigated: (1) improving the routes for CO_2 fixation, (2) modifying complexes that collect energy, and (3) creating a new carbon fixation pathway to reduce energy and CO_2

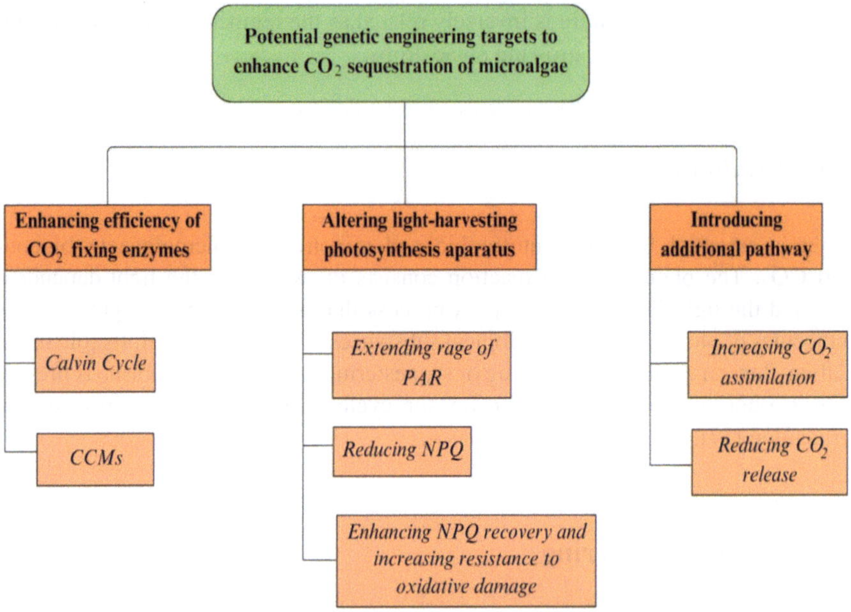

Fig. 1 The overview of approaches to improve CO_2 sequestration of microalgae

emissions. Figure 1 shows a summary of methods that have been proposed to enhance microalgae CO_2 sequestration.

6.1 *Making CO_2-Fixation Routes More Efficient*

One of the most important approaches to improving CO_2 fixation efficiency is to regulate the expression of the key enzymes in the pathways that fix CO_2. The process of carbon dioxide fixing involves two important processes; the process by which organic carbon is fixed to inorganic carbon through the Calvin cycle; and CCMS, which makes it easier for inorganic carbon to be delivered to the Calvin cycle.

6.1.1 Enhancement of Enzymes Essential to the Calvin Cycle

As long as carbon dioxide and oxygen are present, Rubisco can catalyze the ribulose 1,5 bisphosphate reactions, both the photorespiratory route, which releases carbon dioxide, and the Calvin cycle, which fixes carbon dioxide, are given equal metabolic flux [61]. According to Spreitzer and Salvucci [62], Rubisco is an exceedingly sluggish enzyme with an exceptionally low intrinsic affinity for carbon dioxide. While CO_2 levels are typical, Rubisco only uses about a quarter of its catalytic capacity because the concentration of dissolved CO_2 is smaller than the Km, which races, with CO_2, for CO_2 of Rubisco, and the level of is relatively higher. Greater understanding of Rubisco synthesis and assembly, together with the discovery of catalytic alterations in amino acids inside the large subunit (rbcL), has led to the most important advancements in Rubisco engineering in the last several years [63]. It is possible to enhance Rubisco's performance through the use of protein engineering. But there have been few efforts to apply it to other taxa and even fewer to microalgae.

6.1.2 Elevation of CCM Features

The second important class of carbon fixation enzymes are carbonic anhydrases (CAs) that are involved with CCM. The initial evidence of a CCM in green microalgae was found in *C. reinhardtii* [64]. The Ci transport, carboxysomes, and CAs are the three main parts of CCM that can be discussed in terms of their structure and functionality. The process of converting HCO_3 for Rubisco is facilitated by CAs [65]. Results showed that CA activity increased immediately upon cell concentration reduction, and an increase in the photosynthetic CO_2 fixation rate in limiting conditions was transmitted via an escalation. As a result, we know that CA enhances photosynthesis during conditions of high CO_2 production, but reduces photo levels when CO_2 is limiting.

Several studies have examined how various amounts of CO_2 affect the expression of genes associated with CCM [66–69]. In their study, Wang et al. [70] identified several genes that are crucial for the proper functioning of an algal chlorophyll-dependent transport system (CCMS). These genes include those that capture chlorophyll-acids (CAH1, CAH2, CAH3, CAH4, CAH5, and CAH6), transporters of cis–trans ions (HLA3, LCI1, LCIA, CCP1, CCP2, CemA, RHP1, and RHP2), transporters of carbon dioxide (CO_2), a putative methyl transferase (CIA6), and two nuclear transcriptional regulators (CIA5/CCM1 and LCR1). As an example, at low concentrations of CO_2, HLA3 and LCI1 are crucial. The HLA3 protein is an important player in the process of active carbon absorption [71, 72]. It has also been proven that LCI1 acts as a CO_2 channel within the plasma membrane [73].

6.2 *Modifying Light-Harvesting Photosynthetic Apparatus*

There are three approaches to improving energy harvesting complexes: (1) extending the range of radiation that is photosynthetically active [74], (2) lessening the NPQ, or non-photochemical dissipation of excitement, and (3) enhancing antioxidant defenses through NPQ manipulation [75].

6.2.1 Extending Photosynthetically Active Radiation Range

The range of wavelengths in the sun's spectrum that can be harnessed to start photosynthesis is known as photosynthetic active radiation (PAR). The two main subtypes of chlorophyll found in higher plants and green algae are a and b. One current idea is to use routes for the construction of chlorophyll f and d, which are found in some cyanophytes that absorb above 700 nm, to further extend the PAR range in eukaryotic phototrophs [76]. Increasing the overall accessibility of PAR photons by 19% is possible with the insertion of various chlorophylls into higher plants or microalgae. This might result in significantly larger energy for photosynthesis if the PAR range extends to 750 nm [77]. The plant Acaryochloris marina was the first to contain chlorophyll d [78]. Although the exact catalyst or enzyme involved in chlorophyll d production is yet unknown, it has been postulated that the FaRLiP gene cluster is the locus from which Chlorophyll d synthase is encoded [79].

6.2.2 Reducing NPQ

About 600 chlorophylls linked to photosystem I (PSI) and photosystem II (PSII) make up the bulk of an antenna structure found in the majority of algae [80]. To maximize output per unit area, cells are densely packed in industrial-scale growing systems. As a result of self-shading and light weakening in deeper layers of the culture, the excessive concentration of antenna pigments turns out to be harmful [81]. When exposed to more light than is necessary for photosynthesis, surface cells absorb the majority of it, which causes the excess photons to be dissipated through NPQ processes and, likely, photo inhibition. Light would be rapidly attenuated to a level near or below the compensation threshold between photosynthesis and respiration in most cultures due to self-shading caused by high pigment optical density [80]. Minimizing the size of the chlorophyll antenna is one approach to addressing issues,this would lead to less light absorption and more efficient and productive photosynthetic processes. Microalgae biomass increased dramatically, especially in dense cultures, as a consequence of the antenna size reduction in photosystems [82]. Maximizing photosynthetic production while decreasing energy consumption due to NPQ avoidance is possible with a smaller light-harvesting antenna [83]. As a result, photosynthetic productivity and the efficiency of light conversion to biomass can be enhanced in dense cultures. Additionally, it increases the likelihood that cells will be exposed to solar light by preventing the shading effect during mass culture. Assembly of chlorophyll-carotenoid light-harvesting complexes (LHCs) in conjunction with the reaction centers of PSI and PSII is how photosynthetic organisms gather light energy [84]. The size of the chlorophyll antenna can be controlled by controlling the expression of LHC genes, their trafficking into the chloroplast, and their assembly into the membrane.

6.2.3 Manipulating NPQ Mechanisms and Increasing Resistance to Oxidative Damage

A decrease in photosynthetic organisms' NPQ is known to increase crop yields because light continues to dissipate under low light settings after a few minutes of transitioning from high light [85]. Generating cells that can quickly recover from photoprotection could potentially increase carbon uptake, which in turn would improve adaptation to sudden changes in light [86] and biomass productivity [87]. The LHCSR3 protein is also critical to the PSI and PSII NPQ mechanisms in C. reinhardtii [88, 89]. The expression of LHCSR3 was regulated by the HSP70/RBCS2 promoter in order to ensure that the cell had a minimal NPQ capacity. According to Perozeni et al. [90], this fine-tuned NPQ activity increased C. reinhardtii photosynthetic efficiency from 3.4 to 4.4–4.7%. This finding implies that expressing the main regulatory NPQ genes correctly could be a potential way to increase biomass output. Gabilly et al. [91] found that C. reinhardtii's quick NPQ recovery is controlled by the SPA1 and CUL4 genes, which regulate the expression of the PSBS and LHCSR genes in response to high light, blue light, and UV radiation. These genes have the potential to be excellent candidates for optimizing NPQ recovery.

6.3 Introducing Additional Carbon Fixation Pathway to Minimize CO_2 Energy Loss

Scientists have tried to modify microalgae photorespiration pathways in an effort to increase the effectiveness of their photosynthetic processes by replacing the Rubisco enzyme with more effective carboxylases or by finding alternative carbon fixation pathways for the Calvin cycle.

6.3.1 Increasing CO_2 Assimilation

For billions of years, Rubisco has withstood evolutionary pressure without improving its efficiency. Efforts to enhance it have not yielded notable outcomes thus far. Substituting it with other enzymes is a more promising alternative. Given its excellent kinetic and oxygenation properties, the enzyme 6-phosphogluconate dehydrogenase (6PGDH) was suggested as a potential replacement [92]. Though it produces a distinct product, 6-phosphogluconate (6 PG), which is not recycled into the Calvin cycle, 6PGDH could react with Rubisco's substrate as a reductive carboxylase. Despite the fact that this could be a huge obstacle to overcome, Bar-Even outlined methods for recycling 6 PG into the Calvin cycle.

6.3.2 Reducing CO_2 Release

Photosynthesis is inefficient for several reasons, one of which is photorespiration. In contrast to Calvin cycle's carbon fixing, photorespiration releases carbon dioxide. In addition, the assimilation of ATP and decreased ferredoxins is necessary for this process, which results in the emission of ammonia. Furthermore, oxygen oxidizes glycolate, making use of the redox energy. Finally, the conversion of glycerate to 3-phosphoglycerate (3PGA) requires an additional ATP in the PR route. Two molecules of 2-phosphoglycerate (2 PG) were reabsorbed into one 3PGA after 3.5 ATP and 2 NADPH were used [93, 94]. However, chemicals that are part of the anti-oxidative defense system, such as glutathione, glycine, and serine, can be produced during photorespiration [95]. Thus, photorespiration might be both a wasteful and a useful process. Mutants with impaired photorespiration were demonstrated to be conditionally lethal in the Arabidopsis plant. These mutants grew almost identically to their wild-type counterparts in environments with high concentrations of carbon dioxide, which prevented photorespiration. However, these mutants exhibited severe stress symptoms when transferred from environments with high CO_2 levels to those with atmospheric CO_2 levels. When exposed to high levels of CO_2, though, they made a full recovery [96]. It follows that other processes, including as nitrogen absorption, protein synthesis, and amino acid metabolism, would be drastically affected by a reduction in photorespiration [97].

7 Genetic Modification Approach to Improve CO_2 Sequestration by Microalgae

7.1 Random Mutagenesis

Processing random mutagenesis is simple, and massive amounts of organism data are not necessary. Ethyl nitrosourea (ENU), nitrosoguanidine (GTN), ethyl methane sulfonate (EMS), methyl methane sulfate (MMS), N-methyl N-nitrosourea (MNU), and N-ethyl-N-nitrosourea (ENU) are among the chemical mutagens that can cause random mutations [98]. To produce point mutations, like the substitution of A-T for G-C, random mutagenesis by EMS is an effective and encouraging method. Many studies have shown that EMS mutagenesis increases biomass and lipid synthesis in many types of microalgae, which are essential components of biodiesel [99, 100]. One example is a mutant strain of *Candida vulgaris* that showed 45.5% increased biomass productivity when exposed to intense light. This strain was created using EMS mutagenesis [101]. C. vulgaris was mutated twice in an additional intriguing investigation. First, we isolated the mutants that had smaller antennas,second, we picked the mutants that had higher resistance to reactive oxygen species (ROS). Biomass productivity increased by 68% in the mutants. The use of electroshock mutagenesis allowed for the creation of two mutant *Desmodesmus* sp. strains that displayed enhanced biomass and lipid production [102]. Some of the difficulties associated with random mutagenesis include mutant screening. But there are situations where the fluorescence properties can be used to screen. As an example, a 69.3% increase in fucoxanthin was achieved by generating mutants of the diatom P. tricornutum using EMS and N-methyl-N'-nitro-N-nitrosoguanidine (NTG). These mutants were then detected using fluorescence screening [103]. Nuclear radiation mutagenesis has several applications, similar to chemical mutagens. Mutagenesis caused by nuclear radiation mostly impacts DNA molecules through breaking glycosidic bonds or triggering base oxidation. Different locations of DNA were mutated many times due to the 60Co-γ ray mutagenesis [104]. At a high CO_2 level of 15%, chlorella species subjected to nuclear irradiation demonstrated enhanced biomass and CO_2 fixation [105]. Using air bubbling, the mutant increased biomass output by 53.1%. In addition, the biomass production was enhanced to 2.41 g L^{-1} by cultivating the mutant with successively higher CO_2 concentrations up to 15%. In the same way, Spirulina sp. that were mutated using 60Co-γ demonstrated enhanced growth and the ability to fixate CO_2 in a 15% CO_2 environment. The biomass productivity was increased by 310% in this mutant. This mutant's biomass productivity increased by 500% after acclimation under 15% CO_2. While cultivating in a 15% CO_2 environment, researchers found that the mutant cell wall's pore size had increased by 33%, facilitating the direct passage of CO_2 [106]. Additionally, microalgae have had random mutations introduced into them using ultraviolet (UV) light in order to increase their CO_2 utilization [107–109]. As an example, a mutant strain of C. sorokiniana was created that had a significantly smaller antenna. In both laboratory and industrial outdoor cultivations, this mutant exhibited a 30% increase in biomass productivity [110]. When looking to create new mutants, random mutagenesis is a lifesaver. But it's labor-intensive, takes a long time, and could produce mutants whose effects are unclear [111]. As an example, two D. tertiolecta mutant cell lines generated via EMS mutagenesis showed improved photosynthetic efficiency and biomass productivity, but they differed significantly in their capacity to adapt to various light conditions [112]. Transcriptome analysis showed, however, that both mutant cell lines had comparable profiles of disrupted nuclear-encoded photosynthesis genes. More research is needed to understand how this works and to pinpoint which genes cause which phenotypes when exposed to bright light. The following provides more detail on how recent developments in targeted gene editing have led to more accurate results.

7.2 Targeted Genetic Modifications

Microalgal cells can be infected using Agrobacterium, biolistic, electroporation, or agitation with glass beads to introduce DNA [113]. Each of these approaches has its own set of advantages and disadvantages, as well as its own unique set of processes, efficiency, and genomic integration sites [114]. A well-known, easy, and quick technique for transformation is the use of glass beads. However, electroporation has proven to be more effective, despite being more complex and requiring specialized equipment. The cell wall-deficient mutants created using this method outperformed those created using the glass bead method by a factor of 9 to 163 in terms of transformation efficiency in C. reinhardtii [115]. This method could be further improved upon using square-wave electroporation. It is possible to change microalgae that have a cell wall via particle bombardment, which is also known as biokinetics. This method effectively penetrates microalgal cell walls,however, it is limited in its application due to the expensive equipment and procedure-specific parameters that are required. In contrast to electroporation and glass-bead agitation, the biolistic approach is superior for chloroplast transformation [116]. Consequently, picking the right delivery mechanism is crucial for a smooth transition. One of the major obstacles during transformation is the random integration of the introduced gene. The use of homologous recombination transformation, one of several recognized methods for site-specific gene insertion in target cells, helps to decrease random integrations [117, 118]. The transformation efficiency was low, even if the targeted insertion rate may be increased using the homologous recombination technique. In addition, there are additional concerns that need resolving, such as the fact that genetic stability and constant vertical transmission are critical aspects that will impact the changed cell's practical applicability. In addition, the majority of mutant algae express the recombinant gene as they grow, which might have a negative impact on growth owing to the increased metabolic load or the cell-damaging effects of the produced product [119]. Modern gene editing tools can greatly facilitate the process of knocking out or inserting new genes, which is often necessary for rewiring the native metabolic system. Recent advancements in genomic modification techniques have ushered in a new era in genetic editing, allowing for precise manipulation of specific genomic sites. A technique called CRISPR-Cas9 has recently appeared,it consists of CRISPR and the associated protein Cas9. This tool can accurately target a specific location in the genome guided by a design 20 bp RNA sequence [120]. It is possible to conduct functional genome research without making any permanent changes to the genome, precise knockouts [121], and alterations to gene expression [122]. A novel approach to functional genomic screening, base editing using CRISPR, has also recently emerged. In comparison to other gene editing methods, CRISPR is easier to use, more accurate, and less expensive [123]. Furthermore, some researchers think this approach could circumvent the associated problems with GMOs since it does not incorporate any synthetic DNA, unlike GMOs, which can harbor markers and undesirable heterologous genes [124]. Nonetheless, remarkable as it may seem, a study found that gene-editing effectiveness was enhanced by over 30% through the insertion of a huge DNA fragment including luciferase (aph7) as a selectable marker and mVenus as a reporter gene [125]. Jiang et al. [126] were the first to establish gene editing in microalgae using CRISPR technology. Next, a plasmid vector was not needed in the DNA-Free technique [127] since ribonucleoproteins (RNPs) were utilized and a number of knockout and knockin mutations were successfully introduced into C. reinhardtii. The DNA-free method greatly improved upon this technology since it eliminated the cytotoxicity and off-targets that are common in vector expression [127]. Using CRISPR technology, multiple efforts have been made in recent years to increase oleaginous microalgae lipid output. One example is the oleaginous microalgae N. oceanica IMET1, which was engineered to lack the nitrate reductase gene, which plays a role in nitrogen assimilation. This sample developed well in NH_4Cl because the precise five-base deletion caused a frameshift in the NR amino acid sequence, but died out when exposed to a $NaNO_3$-only medium [128]. Additionally, the expression of rfp and PEPC1 in C. reinhardtii was controlled using CRISPR interference (CRISPRi) [129]. The PEPC1 gene encodes a protein that competes with lipid synthesis for the opportunity to divide carbon atoms in various metabolites. The strains with down-regulated CrPEPC1 had lower chlorophyll contents but significantly higher rates of biomass accumulation and lipid synthesis. Table 2 displays a number of reports regarding the use of CRISPR technology for microalgal gene editing. Microalgae gene editing has historically focused on reducing antenna size for the sake of easier phenotypic screening.

8 Conclusions

Carbon dioxide sequestration by microalgae reduces atmospheric carbon and contributes to slowing the rate of global warming. According to the chapter's thorough analysis of microalgae growth parameters, microalgae are both very tolerant and very sensitive to temperature, pH, irradiance, and nutritional conditions; nonetheless, even little changes in the cultivation conditions can significantly affect the product output. Recent progress in the biorefinery-based growth system gives us hope for large-scale carbon sequestration, and microalgae cultivation technologies also play an essential role in large-scale production. Finally, a better understanding of the photosynthetic mechanism, improved growth factors, and technological infrastructure are crucial to the long-term viability and economic viability of large-scale microalgae carbon sequestration. Many investigations have concentrated on photosynthesis-related enzymes, either by over-expressing them or by knocking them off. Shortening the length of an antenna is supposedly one of the successful genetic engineering aims. In addition, with the use of CRISPR technology and a logical design, it is feasible to alter important enzymes like Rubisco. Utilizing the most recent findings and technologies to enhance CO_2 usage in flue gas has the potential to greatly enhance microalgae productivity, leading to broader applications.

Table 2 List of gene editing reports in microalgae using CRISPR technology

Organism	Strategy	Technique of delivery	System	Target and remarks	Ref
C. reinhardtii	Truncating antenna size	Electroporation	Vector construction	First application of CRISPR/ CAS9 in microalgae. Transient expression of Cas9 and sgRNA Genes. Knockout of FKB12 (conferring sensitivity to rapamycin), while continuous CAS9 expression led to cell toxicity and mortality	[126]
C. reinhardtii	Truncating antenna size	Electroporation	CRISPR-Cas9 RNP	Knockout of CpFTSY (photosystem antenna size regulation) and ZEP (catalyzing zeaxanthin epoxidation) led to the continuous production of zeaxanthin and improved photosynthetic productivity	[130]
C. reinhardtii	Truncating antenna size	Electroporation	CRISPR-Cas9 RNP	Knockdown of MAA7 (encodes the beta subunit of tryptophan synthase) and CpSRP43 (encodes chloroplast SRP43 protein) led to truncated antenna size and improved photosynthetic productivity	[131]

(continued)

Table 2 (continued)

Organism	Strategy	Technique of delivery	System	Target and remarks	Ref
C. reinhardtii	Truncating antenna size	Electroporation	CRISPR-Cas9 RNP	FKB12, CpFTSY, CpSRP43, and PHT7 targeted. Single strand DNA used to promote HDR led to the significantly higher efficiency of gene editing	[132]
C. reinhardtii	Truncating antenna size	Electroporation	CRISPR-Cas9 RNP	Knockout of LTD (LHCP translocation defect; involved in chlorophyll assembly) resulted in reduced antenna size	
P. tricornutum	Truncating antenna size	Biolistic	Vector construction	Knockout of chloroplast signal recognition particle 54 (CpSRP54) resulted in truncated antenna size, which led to decreased photosynthetic efficiency	[133]
Coccomyxa sp.	Truncating antenna size	Electroporation	CRISPR-Cas9 RNP	Knockout of the FTSY gene resulted in reduced antenna size	[131]

Declaration of Competing Interest The writers of this paper affirm that they are free from any ties or financial conflicts of interest that may have seemed to impact their work.

References

1. Hansen J, Sato M, Ruedy R, Lacis A, Oinas V (2000) Global warming in the twenty-first century: an alternative scenario. Proc Natl Acad Sci 97(18):9875–9880. https://doi.org/10.1073/pnas.170278997
2. Greenhouse Gas (GHG) (2025b) Emissions and removals. US EPA [WWW Document]. US EPA. https://www.epa.gov/ghgemissions
3. Nduka A (2016) RETRACTED: climate change: is carbon dioxide the culprit? Atmos Clim Sci 06(04):483–488. https://doi.org/10.4236/acs.2016.64038
4. Joint I, Doney SC, Karl DM (2010) Will ocean acidification affect marine microbes? ISME J 5(1):1–7. https://doi.org/10.1038/ismej.2010.79
5. Yang X, Liu L, Yin Z, Wang X, Wang S, Ye Z (2020) Quantifying photosynthetic performance of phytoplankton based on photosynthesis–irradiance response models. Environ Sci Eur 32(1). https://doi.org/10.1186/s12302-020-00306-9
6. Shreyash N, Sonker M, Bajpai S, Tiwary SK, Khan MA, Raj S, Sharma T, Biswas S (2021) The review of carbon capture-storage technologies and developing fuel cells for enhancing utilization. Energies 14(16):4978. https://doi.org/10.3390/en14164978
7. Osman AI, Hefny M, Maksoud MIaA, Elgarahy AM, Rooney DW (2020) Recent advances in carbon capture storage and utilisation technologies: a review. Environ Chem Lett 19(2):797–849. https://doi.org/10.1007/s10311-020-01133-3
8. Chew KW, Yap JY, Show PL, Suan NH, Juan JC, Ling TC, Lee DJ, Chang JS (2017) Microalgae biorefinery: high value products perspectives. Biores Technol 229:53–62. https://doi.org/10.1016/j.biortech.2017.01.006
9. Faried M, Samer M, Abdelsalam E, Yousef R, Attia Y, Ali A (2017) Biodiesel production from microalgae: processes, technologies and recent advancements. Renew Sustain Energy Rev 79:893–913. https://doi.org/10.1016/j.rser.2017.05.199
10. Klinthong W, Yang YH, Huang CH, Tan CS (2015) A review: microalgae and their applications in CO_2 capture and renewable energy. Aerosol Air Qual Res 15(2):712–742. https://doi.org/10.4209/aaqr.2014.11.0299
11. Felipe C, De Souza Michele DRaZ, De Morais Michele G, Bruna DSV, De Morais Etiele G, Jorge AVC (2016) Biofixation of CO_2 on a pilot scale: scaling of the process for industrial application. Afr J Microbiol Res 10(21):768–774. https://doi.org/10.5897/ajmr2015.7641
12. Thomas DM, Mechery J, Paulose SV (2016) Carbon dioxide capture strategies from flue gas using microalgae: a review. Environ Sci Pollut Res 23(17):16926–16940. https://doi.org/10.1007/s11356-016-7158-3
13. Zhu X, Rong J, Chen H, He C, Hu W, Wang Q (2016) An informatics-based analysis of developments to date and prospects for the application of microalgae in the biological sequestration of industrial flue gas. Appl Microbiol Biotechnol 100(5):2073–2082. https://doi.org/10.1007/s00253-015-7277-7
14. Cuellar-Bermudez SP, Garcia-Perez JS, Rittmann BE, Parra-Saldivar R (2015) Photosynthetic bioenergy utilizing CO_2: an approach on flue gases utilization for third generation biofuels. J Clean Prod 98:53–65. https://doi.org/10.1016/j.jclepro.2014.03.034
15. Lam MK, Lee KT, Mohamed AR (2012) Current status and challenges on microalgae-based carbon capture. Int J Greenhouse Gas Control 10:456–469. https://doi.org/10.1016/j.ijggc.2012.07.010
16. Khan SA, Rashmi N, Hussain MZ, Prasad S, Banerjee U (2009) Prospects of biodiesel production from microalgae in India. Renew Sustain Energy Rev 13(9):2361–2372. https://doi.org/10.1016/j.rser.2009.04.005
17. Barrett J, Girr P, Mackinder LC (2021) Pyrenoids: CO_2-fixing phase separated liquid organelles. Biochimica Et Biophysica Acta (BBA) Mol Cell Res 1868(5):118949. https://doi.org/10.1016/j.bbamcr.2021.118949
18. Wang Y, Stessman DJ, Spalding MH (2015) The CO 2 concentrating mechanism and photosynthetic carbon assimilation in limiting CO 2: how Chlamydomonas works against the gradient. Plant J 82(3):429–448. https://doi.org/10.1111/tpj.12829

19. Yen H, Ho S, Chen C, Chang J (2015) CO_2, NOx and SOx removal from flue gas via microalgae cultivation: a critical review. Biotechnol J 10(6):829–839. https://doi.org/10.1002/biot.201400707
20. Larkum AWD, Ross IL, Kruse O, Hankamer B (2012) Selection, breeding and engineering of microalgae for bioenergy and biofuel production. Trends Biotechnol 30(4):198–205. https://doi.org/10.1016/j.tibtech.2011.11.003
21. Raheem A, Prinsen P, Vuppaladadiyam AK, Zhao M, Luque R (2018) A review on sustainable microalgae based biofuel and bioenergy production: recent developments. J Clean Prod 181:42–59. https://doi.org/10.1016/j.jclepro.2018.01.125
22. Arita CEQ, Peebles C, Bradley TH (2015) Scalability of combining microalgae-based biofuels with wastewater facilities: a review. Algal Res 9:160–169. https://doi.org/10.1016/j.algal.2015.03.001
23. Kothari R, Prasad R, Kumar V, Singh D (2013) Production of biodiesel from microalgae Chlamydomonas polypyrenoideum grown on dairy industry wastewater. Biores Technol 144:499–503. https://doi.org/10.1016/j.biortech.2013.06.116
24. Singer AM, Branham M, Hutchins MG, Welker J, Woodard DL, Badurek CA, Ruseva T, Marland E, Marland G (2014) The role of CO_2 emissions from large point sources in emissions totals, responsibility, and policy. Environ Sci Policy 44:190–200. https://doi.org/10.1016/j.envsci.2014.08.001
25. Schubert BA, Jahren AH (2012) The effect of atmospheric CO_2 concentration on carbon isotope fractionation in C3 land plants. Geochim Cosmochim Acta 96:29–43. https://doi.org/10.1016/j.gca.2012.08.003
26. Liu M, Wu J, Zhu X, He H, Jia W, Xiang W (2015) Evolution and variation of atmospheric carbon dioxide concentration over terrestrial ecosystems as derived from eddy covariance measurements. Atmos Environ 114:75–82. https://doi.org/10.1016/j.atmosenv.2015.05.026
27. Luo X, Wang J, Dooner M, Clarke J (2015) Overview of current development in electrical energy storage technologies and the application potential in power system operation. Appl Energy 137:511–536. https://doi.org/10.1016/j.apenergy.2014.09.081
28. He JK (2015) China's INDC and non-fossil energy development. Adv Clim Chang Res 6(3–4):210–215. https://doi.org/10.1016/j.accre.2015.11.007
29. Miller L, Carriveau R (2019) Energy demand curve variables–an overview of individual and systemic effects. Sustain Energy Technol Assess 35:172–179. https://doi.org/10.1016/j.seta.2019.07.006
30. Nocito F, Dibenedetto A (2020) Atmospheric CO_2 mitigation technologies: carbon capture utilization and storage. Curr Opin Green Sustain Chem 21:34–43. https://doi.org/10.1016/j.cogsc.2019.10.002
31. Gurney KR, Mendoza DL, Zhou Y, Fischer ML, Miller CC, Geethakumar S, De La Rue Du Can S (2009) High resolution fossil fuel combustion CO_2 emission fluxes for the United States. Environ Sci Technol 43(14):5535–5541.https://doi.org/10.1021/es900806c
32. Worrell E, Price L, Martin N, Hendriks C, Meida LO (2001) Carbon dioxide emissions from the global cement industry. Annu Rev Energy Env 26(1):303–329. https://doi.org/10.1146/annurev.energy.26.1.303
33. Lu I, Lewis C, Lin SJ (2009) The forecast of motor vehicle, energy demand and CO_2 emission from Taiwan's road transportation sector. Energy Policy 37(8):2952–2961. https://doi.org/10.1016/j.enpol.2009.03.039
34. Davis SJ, Caldeira K, Matthews HD (2010) Future CO 2 emissions and climate change from existing energy infrastructure. Science 329(5997):1330–1333. https://doi.org/10.1126/science.1188566
35. Cai Q, Yan X, Li Y, Wang L (2018) Global patterns of human and livestock respiration. Sci Rep 8(1). https://doi.org/10.1038/s41598-018-27631-7
36. Oertel C, Matschullat J, Zurba K, Zimmermann F, Erasmi S (2016) Greenhouse gas emissions from soils—a review. Geochemistry 76:327–352. https://doi.org/10.1016/j.chemer.2016.04.002

37. Greenwell HC, Laurens LML, Shields RJ, Lovitt RW, Flynn KJ (2009) Placing microalgae on the biofuels priority list: a review of the technological challenges. J R Soc Interface 7:703–726. https://doi.org/10.1098/rsif.2009.0322
38. Brennan L, Owende P (2010) Biofuels from microalgae—a review of technologies for production, processing, and extractions of biofuels and co-products. Renew Sustain Energy Rev 14:557–577. https://doi.org/10.1016/j.rser.2009.10.009
39. Bhola V, Swalaha F, Kumar RR, Singh M, Bux F (2014) Overview of the potential of microalgae for CO_2 sequestration. Int J Environ Sci Technol 11:2103–2118. https://doi.org/10.1007/s13762-013-0487-6
40. Huertas IE, Colman B, Espie GS, Lubian LM (2000) Active transport of CO_2 by three species of marine microalgae. J Phycol 36:314–320. https://doi.org/10.1046/j.1529-8817.2000.99142.x
41. Vaz-Pinto F, Rodil IF, Mineur F, Olabarria C (2014) Understanding biological invasions by Seaweeds. In: CRC Press eBooks, pp 148–185. https://doi.org/10.1201/b17540-6
42. Saifuddin N, Aisswarya K, Juan Y, Priatharsi P (2015) Sequestration of high carbon dioxide concentration for induction of lipids in microalgae for biodiesel production. J Appl Sci 15:1045–1058.https://doi.org/10.3923/jas.2015.1045.1058
43. Alami AH, Faraj M, Aokal K, Hawili AA, Tawalbeh M, Zhang D (2020) Investigating various permutations of copper iodide/FeCu tandem materials as electrodes for dye-sensitized solar cells with a natural dye. Nanomaterials 10:784. https://doi.org/10.3390/nano10040784
44. Cheah WY, Show PL, Chang J-S, Ling TC, Juan JC (2015) Biosequestration of atmospheric CO_2 and flue gas-containing CO_2 by microalgae. Biores Technol 184:190–201. https://doi.org/10.1016/j.biortech.2014.11.026
45. Ghorbani A, Rahimpour HR, Ghasemi Y, Zoughi S, Rahimpour MR (2014) A review of carbon capture and sequestration in Iran: microalgal biofixation potential in Iran. Renew Sustain Energy Rev 35:73–100. https://doi.org/10.1016/j.rser.2014.03.013
46. Jiang Y, Zhang W, Wang J, Chen Y, Shen S, Liu T (2013) Utilization of simulated flue gas for cultivation of Scenedesmus dimorphus. Biores Technol 128:359–364. https://doi.org/10.1016/j.biortech.2012.10.119
47. Jiang L, Luo S, Fan X, Yang Z, Guo R (2011) Biomass and lipid production of marine microalgae using municipal wastewater and high concentration of CO_2. Appl Energy 88:3336–3341. https://doi.org/10.1016/j.apenergy.2011.03.043
48. Morales-Sánchez D, Martinez-Rodriguez OA, Kyndt J, Martinez A (2014) Heterotrophic growth of microalgae: metabolic aspects. World J Microbiol Biotechnol 31:1–9. https://doi.org/10.1007/s11274-014-1773-2
49. Solovchenko A, Khozin-Goldberg I (2013) High-CO_2 tolerance in microalgae: possible mechanisms and implications for biotechnology and bioremediation. Biotech Lett 35:1745–1752. https://doi.org/10.1007/s10529-013-1274-7
50. Singh SP, Singh P (2014) Effect of CO_2 concentration on algal growth: a review. Renew Sustain Energy Rev 38:172–179. https://doi.org/10.1016/j.rser.2014.05.043
51. Azov Y (1982) Effect of pH on inorganic carbon uptake in algal cultures. Appl Environ Microbiol 43:1300–1306. https://doi.org/10.1128/aem.43.6.1300-1306.1982
52. Yin D, Wang Z, Wen X, Ding Y, Hou X, Geng Y, Li Y (2019) Effects of carbon concentration, pH, and bubbling depth on carbon dioxide absorption ratio in microalgae medium. Environ Sci Pollut Res 26:32902–32910. https://doi.org/10.1007/s11356-019-06287-4
53. Jabri HA, Taleb A, Touchard R, Saadaoui I, Goetz V, Pruvost J (2021) Cultivating microalgae in desert conditions: evaluation of the effect of light-temperature summer conditions on the growth and metabolism of nannochloropsis QU130. Appl Sci 11:3799. https://doi.org/10.3390/app11093799
54. Ördög V, Stirk WA, Bálint P, Aremu AO, Okem A, Lovász C, Molnár Z, Van Staden J (2016) Effect of temperature and nitrogen concentration on lipid productivity and fatty acid composition in three Chlorella strains. Algal Res 16:141–149. https://doi.org/10.1016/j.algal.2016.03.001

55. Pueyo JJ, Alfonso M, Andrés C, Picorel R (2002) Increased tolerance to thermal inactivation of oxygen evolution in spinach Photosystem II membranes by substitution of the extrinsic 33-kDa protein by its homologue from a thermophilic cyanobacterium. Biochimica Et Biophysica Acta (BBA) Bioenergetics 1554:29–35. https://doi.org/10.1016/s0005-2728(02)00208-6
56. Liang F, Lindberg P, Lindblad P (2018) Engineering photoautotrophic carbon fixation for enhanced growth and productivity. Sustain Energy Fuels 2:2583–2600. https://doi.org/10.1039/c8se00281a
57. Morales-Sánchez D, Martinez-Rodriguez OA, Kyndt J, Martinez A (2014) Heterotrophic growth of microalgae: metabolic aspects. World J Microbiol Biotechnol 31(1):1–9. https://doi.org/10.1007/s11274-014-1773-2
58. Hlavova M, Turoczy Z, Bisova K (2015) Improving microalgae for biotechnology—from genetics to synthetic biology. Biotechnol Adv 33:1194–1203. https://doi.org/10.1016/j.biotechadv.2015.01.009
59. Wei L, Wang Q, Xin Y, Lu Y, Xu J (2017) Enhancing photosynthetic biomass productivity of industrial oleaginous microalgae by overexpression of RuBisCO activase. Algal Res 27:366–375. https://doi.org/10.1016/j.algal.2017.07.023
60. Paul PEV, Sangeetha V, Deepika RG (2019) Emerging trends in the industrial production of chemical products by microorganisms. In Elsevier eBooks, pp 107–125. https://doi.org/10.1016/b978-0-12-816328-3.00009-x
61. Ogren WL (2003) Affixing the O to Rubisco: discovering the source of photorespiratory glycolate and its regulation. Photosynth Res 76:53–63. https://doi.org/10.1023/a:1024913925002
62. Spreitzer RJ, Salvucci ME (2002) RUBISCO: structure, regulatory interactions, and possibilities for a better enzyme. Annu Rev Plant Biol 53:449–475. https://doi.org/10.1146/annurev.arplant.53.100301.135233
63. Sharwood RE (2016) Engineering chloroplasts to improve Rubisco catalysis: prospects for translating improvements into food and fiber crops. New Phytol 213:494–510. https://doi.org/10.1111/nph.14351
64. Badger MR, Kaplan A, Berry JA (1980) Internal inorganic carbon pool of Chlamydomonas reinhardtii. Plant Physiol 66:407–413. https://doi.org/10.1104/pp.66.3.407
65. Matsuda Y, Colman B (1995) Induction of CO_2 and bicarbonate transport in the Green Alga Chlorella ellipsoidea (I. Time course of induction of the two systems). Plant Physiol 108:247–252. https://doi.org/10.1104/pp.108.1.247
66. Sun X, Shen J, Bai F, Xu N (2016) Transcriptome profiling of the microalga Chlorella pyrenoidosa in response to different carbon dioxide concentrations. Mar Genomics 29:81–87. https://doi.org/10.1016/j.margen.2016.05.002
67. Memmola F, Mukherjee B, Moroney JV, Giordano M (2014) Carbon allocation and element composition in four Chlamydomonas mutants defective in genes related to the CO_2 concentrating mechanism. Photosynth Res 121:201–211. https://doi.org/10.1007/s11120-014-0005-9
68. Huang Y, Cheng J, Lu H, He Y, Zhou J, Cen K (2017) Transcriptome and key genes expression related to carbon fixation pathways in Chlorella PY-ZU1 cells and their growth under high concentrations of CO2. Biotechnol Biofuels 10.https://doi.org/10.1186/s13068-017-0868-z
69. Fan J, Xu H, Li Y (2016) Transcriptome-based global analysis of gene expression in response to carbon dioxide deprivation in the green algae Chlorella pyrenoidosa. Algal Res 16:12–19. https://doi.org/10.1016/j.algal.2016.02.032
70. Wang Y, Stessman DJ, Spalding MH (2015) The CO 2 concentrating mechanism and photosynthetic carbon assimilation in limiting CO 2: how Chlamydomonas works against the gradient. Plant J 82:429–448. https://doi.org/10.1111/tpj.12829
71. Yamano T, Sato E, Iguchi H, Fukuda Y, Fukuzawa H (2015) Characterization of cooperative bicarbonate uptake into chloroplast stroma in the green alga Chlamydomonas reinhardtii. Proc Natl Acad Sci 112:7315–7320. https://doi.org/10.1073/pnas.1501659112
72. Duanmu D, Miller AR, Horken KM, Weeks DP, Spalding MH (2009) Knockdown of limiting-CO 2–induced gene HLA3 decreases HCO 3-transport and photosynthetic Ci affinity in

Chlamydomonas reinhardtii. Proc Natl Acad Sci 106:5990–5995. https://doi.org/10.1073/pnas.0812885106

73. Kono A, Chou T, Radhakrishnan A, Bolla JR, Sankar K, Shome S, Su C, Jernigan RL, Robinson CV, Yu EW, Spalding MH (2020) Structure and function of LCI1: a plasma membrane CO_2 channel in the Chlamydomonas CO2 concentrating mechanism. Plant J 102:1107–1126. https://doi.org/10.1111/tpj.14745
74. Blankenship RE, Tiede DM, Barber J, Brudvig GW, Fleming G, Ghirardi M, Gunner MR, Junge W, Kramer DM, Melis A, Moore TA, Moser CC, Nocera DG, Nozik AJ, Ort DR, Parson WW, Prince RC, Sayre RT (2011) Comparing photosynthetic and photovoltaic efficiencies and recognizing the potential for improvement. Science 332:805–809. https://doi.org/10.1126/science.1200165
75. Dall'Osto L, Cazzaniga S, Guardini Z, Barera S, Benedetti M, Mannino G, Maffei ME, Bassi R (2019) Combined resistance to oxidative stress and reduced antenna size enhance light-to-biomass conversion efficiency in Chlorella vulgaris cultures. Biotechnol Biofuels 12. https://doi.org/10.1186/s13068-019-1566-9
76. Wobbe L, Remacle C (2015) Improving the sunlight-to-biomass conversion efficiency in microalgal biofactories. J Biotechnol 201:28–42. https://doi.org/10.1016/j.jbiotec.2014.08.021
77. Blankenship RE, Chen M (2013) Spectral expansion and antenna reduction can enhance photosynthesis for energy production. Curr Opin Chem Biol 17:457–461. https://doi.org/10.1016/j.cbpa.2013.03.031
78. Miyashita H, Ikemoto H, Kurano N (1996) Chlorophyll d as a major pigment. Nature 383:402. https://doi.org/10.1038/383402a0
79. Ho M-Y, Bryant DA (2019) Global transcriptional profiling of the cyanobacterium chlorogloeopsis fritschii PCC 9212 in far-red light: insights into the regulation of Chlorophyll d synthesis. Front Microbiol 10.https://doi.org/10.3389/fmicb.2019.00465
80. Melis A (2009) Solar energy conversion efficiencies in photosynthesis: minimizing the chlorophyll antennae to maximize efficiency. Plant Sci 177:272–280. https://doi.org/10.1016/j.plantsci.2009.06.005
81. Abomohra AE-F, El-Sheekh M, Hanelt D (2014) Pilot cultivation of the chlorophyte microalga Scenedesmus obliquus as a promising feedstock for biofuel. Biomass Bioenerg 64:237–244. https://doi.org/10.1016/j.biombioe.2014.03.049
82. Mussgnug JH, Thomas-Hall S, Rupprecht J, Foo A, Klassen V, McDowall A, Schenk PM, Kruse O, Hankamer B (2007) Engineering photosynthetic light capture: impacts on improved solar energy to biomass conversion. Plant Biotechnol J 5:802–814. https://doi.org/10.1111/j.1467-7652.2007.00285.x
83. De Mooij T, Janssen M, Cerezo-Chinarro O, Mussgnug JH, Kruse O, Ballottari M, Bassi R, Bujaldon S, Wollman F-A, Wijffels RH (2014) Antenna size reduction as a strategy to increase biomass productivity: a great potential not yet realized. J Appl Phycol 27:1063–1077. https://doi.org/10.1007/s10811-014-0427-y
84. Pan X, Cao P, Su X, Liu Z, Li M (2020) Structural analysis and comparison of light-harvesting complexes I and II. Biochimica Et Biophysica Acta (BBA)-Bioenergetics 1861:148038. https://doi.org/10.1016/j.bbabio.2019.06.010
85. Kromdijk J, Glowacka K, Leonelli L, Gabilly ST, Iwai M, Niyogi KK, Long SP (2016) Improving photosynthesis and crop productivity by accelerating recovery from photoprotection. Science 354:857–861. https://doi.org/10.1126/science.aai8878
86. KüLheim C, Ågren J, Jansson S (2002) Rapid regulation of light harvesting and plant fitness in the field. Science 297:91–93. https://doi.org/10.1126/science.1072359
87. Zhu X-G, Ort DR, Whitmarsh J, Long SP (2004) The slow reversibility of photosystem II thermal energy dissipation on transfer from high to low light may cause large losses in carbon gain by crop canopies: a theoretical analysis. J Exp Bot 55:1167–1175. https://doi.org/10.1093/jxb/erh141
88. Perozeni F, Cazzaniga S, Ballottari M (2019) In vitro and in vivo investigation of chlorophyll binding sites involved in non-photochemical quenching in Chlamydomonas reinhardtii. Plant Cell Environ 42:2522–2535. https://doi.org/10.1111/pce.13566

89. Girolomoni L, Cazzaniga S, Pinnola A, Perozeni F, Ballottari M, Bassi R (2019) LHCSR3 is a nonphotochemical quencher of both photosystems in Chlamydomonas reinhardtii. Proc Natl Acad Sci 116:4212–4217. https://doi.org/10.1073/pnas.1809812116
90. Perozeni F, Stella G, Ballottari M (2018) LHCSR expression under HSP70/RBCS2 promoter as a strategy to increase productivity in microalgae. Int J Mol Sci 19:155. https://doi.org/10.3390/ijms19010155
91. Gabilly ST, Baker CR, Wakao S, Crisanto T, Guan K, Bi K, Guiet E, Guadagno CR, Niyogi KK (2019) Regulation of photoprotection gene expression in Chlamydomonas by a putative E3 ubiquitin ligase complex and a homolog of Constans. Proc Natl Acad Sci 116:17556–17562. https://doi.org/10.1073/pnas.1821689116
92. Bar-Even A (2018) Daring metabolic designs for enhanced plant carbon fixation. Plant Sci 273:71–83. https://doi.org/10.1016/j.plantsci.2017.12.007
93. Weber APM, Bar-Even A (2019) Update: improving the efficiency of photosynthetic carbon reactions. Plant Physiol 179:803–812. https://doi.org/10.1104/pp.18.01521
94. South PF, Cavanagh AP, Liu HW, Ort DR (2019) Synthetic glycolate metabolism pathways stimulate crop growth and productivity in the field. Science 363.https://doi.org/10.1126/science.aat9077
95. Lin T, Rao M, Lu H, Chiou C, Lin S, Chao H, Zheng Z, Cheng H, Lee T (2017) A role for glutathione reductase and glutathione in the tolerance of Chlamydomonas reinhardtii to photo-oxidative stress. Physiol Plant 162:35–48. https://doi.org/10.1111/ppl.12622
96. Somerville CR, Ogren WL (1979) A phosphoglycolate phosphatase-deficient mutant of Arabidopsis. Nature 280:833–836. https://doi.org/10.1038/280833a0
97. Wingler A, Lea PJ, Quick WP, Leegood RC (2000) Photorespiration: metabolic pathways and their role in stress protection. Philos Trans Royal Soci B Biol Sci 355:1517–1529. https://doi.org/10.1098/rstb.2000.0712
98. Moha-León JD, Pérez-Legaspi IA, Ortega-Clemente LA, Rubio-Franchini I, Ríos-Leal E (2018) Improving the lipid content of Nannochloropsis oculata by a mutation-selection program using UV radiation and quizalofop. J Appl Phycol 31:191–199. https://doi.org/10.1007/s10811-018-1568-1
99. Kumar SD, Sojin K, Santhanam P, Dhanalakshmi B, Latha S, Park MS, Kim M-K (2018) Triggering of fatty acids on Tetraselmis sp. by ethyl methanesulfonate mutagenic treatment. Bioresour Technol Rep 2:21–28. https://doi.org/10.1016/j.biteb.2018.04.001
100. Tanadul O-U-M, Noochanong W, Jirakranwong P, Chanprame S (2018) EMS-induced mutation followed by quizalofop-screening increased lipid productivity in Chlorella sp. Bioprocess Biosyst Eng 41:613–619. https://doi.org/10.1007/s00449-018-1896-1
101. Zhang Y, He M, Zou S, Fei C, Yan Y, Zheng S, Rajper AA, Wang C (2016) Breeding of high biomass and lipid producing Desmodesmus sp. by Ethylmethane sulfonate-induced mutation. Biores Technol 207:268–275. https://doi.org/10.1016/j.biortech.2016.01.120
102. Shin W-S, Lee B, Jeong B-R, Chang YK, Kwon J-H (2016) Truncated light-harvesting chlorophyll antenna size in Chlorella vulgaris improves biomass productivity. J Appl Phycol 28:3193–3202. https://doi.org/10.1007/s10811-016-0874-8
103. Yi Z, Su Y, Xu M, Bergmann A, Ingthorsson S, Rolfsson O, Salehi-Ashtiani K, Brynjolfsson S, Fu W (2018) Chemical mutagenesis and fluorescence-based high-throughput screening for enhanced accumulation of carotenoids in a model marine diatom phaeodactylum tricornutum. Mar Drugs 16:272. https://doi.org/10.3390/md16080272
104. Wang W, Wei T, Fan J, Yi J, Li Y, Wan M, Wang J, Bai W (2018) Repeated mutagenic effects of 60Co-γ irradiation coupled with high-throughput screening improves lipid accumulation in mutant strains of the microalgae Chlorella pyrenoidosa as a feedstock for bioenergy. Algal Res 33:71–77. https://doi.org/10.1016/j.algal.2018.04.022
105. Cheng J, Huang Y, Feng J, Sun J, Zhou J, Cen K (2013) Mutate Chlorella sp. by nuclear irradiation to fix high concentrations of CO_2. Biores Technol 136:496–501. https://doi.org/10.1016/j.biortech.2013.03.072
106. Cheng J, Lu H, He X, Yang W, Zhou J, Cen K (2017) Mutation of Spirulina sp. by nuclear irradiation to improve growth rate under 15% carbon dioxide in flue gas. Biores Technol 238:650–656. https://doi.org/10.1016/j.biortech.2017.04.107

107. Liu S, Zhao Y, Liu L, Ao X, Ma L, Wu M, Ma F (2015) Improving cell growth and lipid accumulation in green microalgae chlorella sp. via UV irradiation. Appl Biochem Biotechnol 175:3507–3518. https://doi.org/10.1007/s12010-015-1521-6
108. Sivaramakrishnan R, Incharoensakdi A (2017) Enhancement of lipid production in Scenedesmus sp. by UV mutagenesis and hydrogen peroxide treatment. Biores Technol 235:366–370. https://doi.org/10.1016/j.biortech.2017.03.102
109. Beacham TA, Macia VM, Rooks P, White DA, Ali ST (2015) Altered lipid accumulation in Nannochloropsis salina CCAP849/3 following EMS and UV induced mutagenesis. Biotechnol Rep 7:87–94. https://doi.org/10.1016/j.btre.2015.05.007
110. Cazzaniga S, Dall'Osto L, Szaub J, Scibilia L, Ballottari M, Purton S, Bassi R (2014) Domestication of the green alga Chlorella sorokiniana: reduction of antenna size improves light-use efficiency in a photobioreactor. Biotechnol Biofuels 7. https://doi.org/10.1186/s13068-014-0157-z
111. Perin G, Bellan A, Segalla A, Meneghesso A, Alboresi A, Morosinotto T (2015) Generation of random mutants to improve light-use efficiency of Nannochloropsis gaditana cultures for biofuel production. Biotechnol Biofuels 8.https://doi.org/10.1186/s13068-015-0337-5
112. Johansson SA, Stephenson PG, Edwards RJ, Yoshida K, Moore CM, Terauchi R, Zubkov MV, Terry MJ, Bibby TS (2020) Isolation and molecular characterisation of Dunaliella tertiolecta with truncated light-harvesting antenna for enhanced photosynthetic efficiency. Algal Res 48:101917. https://doi.org/10.1016/j.algal.2020.101917
113. Ng I, Tan S, Kao P, Chang Y, Chang J (2017) Recent developments on genetic engineering of microalgae for biofuels and bio-based chemicals. Biotechnol J 12.https://doi.org/10.1002/biot.201600644
114. Banerjee C, Singh PK, Shukla P (2016) Microalgal bioengineering for sustainable energy development: recent transgenesis and metabolic engineering strategies. Biotechnol J 11:303–314. https://doi.org/10.1002/biot.201500284
115. Shimogawara K, Fujiwara S, Grossman A, Usuda H (1998) High-efficiency transformation of chlamydomonas reinhardtii by electroporation. Genetics 148:1821–1828. https://doi.org/10.1093/genetics/148.4.1821
116. Economou C, Wannathong T, Szaub J, Purton S (2014) A simple, low-cost method for chloroplast transformation of the green alga Chlamydomonas reinhardtii. Methods Mol Biol 401–411.https://doi.org/10.1007/978-1-62703-995-6_27
117. Sodeinde OA, Kindle KL (1993) Homologous recombination in the nuclear genome of Chlamydomonas reinhardtii. Proc Natl Acad Sci 90:9199–9203. https://doi.org/10.1073/pnas.90.19.9199
118. Zorin B, Lu Y, Sizova I, Hegemann P (2009) Nuclear gene targeting in Chlamydomonas as exemplified by disruption of the PHOT gene. Gene 432:91–96. https://doi.org/10.1016/j.gene.2008.11.028
119. Charoonnart P, Purton S, Saksmerprome V (2018) Applications of microalgal biotechnology for disease control in aquaculture. Biology 7:24. https://doi.org/10.3390/biology7020024
120. Jinek M, Chylinski K, Fonfara I, Hauer M, Doudna JA, Charpentier E (2012) A programmable dual-RNA–guided DNA endonuclease in adaptive bacterial immunity. Science 337:816–821. https://doi.org/10.1126/science.1225829
121. Shalem O, Sanjana NE, Zhang F (2015) High-throughput functional genomics using CRISPR–Cas9. Nat Rev Genet 16:299–311. https://doi.org/10.1038/nrg3899
122. Dominguez AA, Lim WA, Qi LS (2015) Beyond editing: repurposing CRISPR–Cas9 for precision genome regulation and interrogation. Nat Rev Mol Cell Biol 17:5–15. https://doi.org/10.1038/nrm.2015.2
123. Conlan B, Whitney S (2017) Preparing rubisco for a tune up. Nat Plants 4:12–13. https://doi.org/10.1038/s41477-017-0089-2
124. Kim J, Kim J-S (2016) Bypassing GMO regulations with CRISPR gene editing. Nat Biotechnol 34:1014–1015. https://doi.org/10.1038/nbt.3680
125. Kim J, Lee S, Baek K, Jin E (2020) Site-specific gene knock-out and on-site heterologous gene overexpression in Chlamydomonas reinhardtii via a CRISPR-Cas9-mediated knock-in method. Front Plant Sci 11.https://doi.org/10.3389/fpls.2020.00306

126. Jiang W, Brueggeman AJ, Horken KM, Plucinak TM, Weeks DP (2014) Successful transient expression of Cas9 and single guide RNA genes in chlamydomonas reinhardtii. Eukaryot Cell 13:1465–1469. https://doi.org/10.1128/ec.00213-14
127. Shin S-E, Lim J-M, Koh HG, Kim EK, Kang NK, Jeon S, Kwon S, Shin W-S, Lee B, Hwangbo K, Kim J, Ye SH, Yun J-Y, Seo H, Oh H-M, Kim K-J, Kim J-S, Jeong W-J, Chang YK, Jeong B-R (2016) CRISPR/Cas9-induced knockout and knock-in mutations in Chlamydomonas reinhardtii. Sci Rep 6.https://doi.org/10.1038/srep27810
128. Wang Q, Lu Y, Xin Y, Wei L, Huang S, Xu J (2016) Genome editing of model oleaginous microalgaeNannochloropsisspp. by CRISPR/Cas9. Plant J 88:1071–1081. https://doi.org/10.1111/tpj.13307
129. Kao P-H, Ng I-S (2017) CRISPRi mediated phosphoenolpyruvate carboxylase regulation to enhance the production of lipid in Chlamydomonas reinhardtii. Biores Technol 245:1527–1537. https://doi.org/10.1016/j.biortech.2017.04.111
130. Baek K, Kim DH, Jeong J, Sim SJ, Melis A, Kim J-S, Jin E, Bae S (2016) DNA-free two-gene knockout in Chlamydomonas reinhardtii via CRISPR-Cas9 ribonucleoproteins. Sci Rep 6.https://doi.org/10.1038/srep30620
131. Yoshimitsu Y, Abe J, Harayama S (2018) Cas9-guide RNA ribonucleoprotein-induced genome editing in the industrial green alga Coccomyxa sp. strain KJ. Biotechnol Biofuels 11. https://doi.org/10.1186/s13068-018-1327-1
132. Ferenczi A, Pyott DE, Xipnitou A, Molnar A (2017) Efficient targeted DNA editing and replacement in Chlamydomonas reinhardtii using Cpf1 ribonucleoproteins and single-stranded DNA. Proc Natl Acad Sci 114:13567–13572. https://doi.org/10.1073/pnas.1710597114
133. Nymark M, Sharma AK, Sparstad T, Bones AM, Winge P (2016) A CRISPR/Cas9 system adapted for gene editing in marine algae. Sci Rep 6.https://doi.org/10.1038/srep24951
134. Singh UB, Ahluwalia AS (2012) Microalgae: a promising tool for carbon sequestration. Mitig Adapt Strat Glob Change 18:73–95. https://doi.org/10.1007/s11027-012-9393-3
135. Gilbert LA, Larson MH, Morsut L, Liu Z, Brar GA, Torres SE, Stern-Ginossar N, Brandman O, Whitehead EH, Doudna JA, Lim WA, Weissman JS, Qi LS (2013) CRISPR-mediated modular RNA-guided regulation of transcription in eukaryotes. Cell 154:442–451. https://doi.org/10.1016/j.cell.2013.06.044

MIX
Papier aus verantwortungsvollen Quellen
Paper from responsible sources
FSC® C105338

If you have any concerns about our products,
you can contact us on
ProductSafety@springernature.com

In case Publisher is established outside the EU,
the EU authorized representative is:
Springer Nature Customer Service Center GmbH
Europaplatz 3, 69115 Heidelberg, Germany

Printed by Libri Plureos GmbH
in Hamburg, Germany